Pla

The Republic. The Essential Philosophical Dialogue

This is a FLAME TREE book

FLAME TREE PUBLISHING
6 Melbray Mews, Fulham,
London SW6 3NS, United Kingdom
www.flametreepublishing.com

This edition is an abridged version of the Benjamin Jowett translation into English, first published in 1871.

24 26 28 27 25
1 3 5 7 9 10 8 6 4 2

ISBN: 978-1-80417-793-8

Cover and pattern art was created by Flame Tree Studio.

A copy of the CIP data for this book is available from the British Library.

Designed and created in the UK | Printed and bound in China

Plato
The Republic. The Essential Philosophical Dialogue

A New Introduction by Professor Harold Tarrant

Series Foreword by Professor Marika Taylor

FLAME TREE CONCISE EDITIONS

Contents

Series Foreword by Professor Marika Taylor **6**

A New Introduction by Harold Tarrant **8**

This Translation and Abridgement 24

The Republic 27

Book I **28**

Omitted – See Introduction.

Book II **29**

The Nature of Justice; The Ring of Gyges; The Rewards and Punishments of Another Life; Justice in the State; The Division of Labour; War; Education; The Falsehoods of the Poets.

Book III **71**

The Reform of Mythology; Epic and Dramatic Poetry; Imitations in Art; Music; Physical Exercise and Health; Selection of Guardians and Rulers.

Book IV **114**

Wealth and Poverty; Education and Lawlessness; Virtues; The Nature of Courage; The Just Man and the Just State; The Criterion of Truth; The Opposition of Desire and Reason; The Principles of the Soul and State.

Book V **154**

Population; Women; Children; Property; Distinctions of Heroes; Behaviour to Enemies; The True Philosopher.

Book VI **202**

The Qualities of Philosophic Nature; The Idea and Nature of Good; The Analogy of the Sun.

Book VII **215**

The Allegory of the Cave; Turning the Mind Upwards; Dialectic and the Cognition of the Good; The Timetable for Education in Mathematics and Dialectic.

Book VIII **229**

Omitted – See Introduction.

Book IX **230**

An Ideal Image of the Soul; A Nobler Purpose and the Paradigm in the Heavens.

Book X **237**

The Immortality of the Soul; The Myth of Er; Conclusion.

Biographies **254**

Series Foreword

Science has always been a source of fascination, pushing the frontiers of knowledge. This series of books spans some of the most important scientific developments of all time. From the origins of life to the nature of the universe, remarkable achievements through the ages are showcased, as well as the limitless potential of human curiosity.

In this series we can follow the evolution of science and philosophy from ancient times through to the modern era. The works of Plato, Ptolemy and Aristotle cover a wide range of topics related to the natural world, including physics, metaphysics and biology. Aristotle's work is particularly notable for its depth and breadth as well as for its immense influence on the development of scientific and philosophical thought in the Western world.

Copernicus and Newton were also great polymaths. Copernicus is best known for his work on the heliocentric nature of the solar system as presented in *On the Revolutions of the*

Heavenly Spheres but he also made substantial contributions to mathematics and medicine. Newton's *Principia* is recognized as one of the most important scientific works ever written. His theories of motion and gravity laid the foundations for modern physics and revolutionized our understanding of the 'Nature' of the world. Within mathematics, Newton developed calculus, as well as algebra, trigonometry and geometry.

The *Principia* was originally read primarily by intellectuals. By contrast, Darwin's work *On the Origin of Species* immediately caused a significant stir in both scientific and public circles, particularly through its challenge to traditional beliefs. Darwin's theory of evolution was debated and refined, gradually becoming a foundational concept of modern biology.

At the beginning of the twentieth century, physics experienced two major paradigm shifts: relativity and quantum theory. Einstein's *Special and General Relativity* fundamentally changed our understanding of space, time and gravity. Bohr and Planck explain the early development of quantum theory, starting from Planck's revolutionary concept of quanta. In the same period, pioneering new fields of research were established all across science, including Freud's renowned psychoanalysis.

Readers of this series can step back in time to explore how the foundations of science and culture are described by its pioneers. The transformative nature of these works and their profound impacts on society can only inspire a sense of awe and wonder.

Professor Marika Taylor
University of Southampton

A New Introduction

The Centrality of Plato's *Republic*

Plato is usually thought to have been born in 428/7 BCE, but D. Nails (*The People of Plato*, 2003, 245–7) argues strongly for 424/3. He may be seen as the central figure in Greek philosophy, having himself been part of Socrates' intellectual circle for the last few years of that figure's life (to 399 BCE), before teaching the young Aristotle for rather longer towards the end of his own life, which concluded in 348/7 BCE. It is likewise arguable that the *Republic* was the central work of his own career, for it is generally agreed to belong to his 'middle period', later than a series of much shorter works that further explored themes in Socratic philosophy, and before several challenging but rather less attractive works, most of which no longer make a pretence of following Socrates.

The *Republic* is also in a sense the central 'political' work, for the bulk of the *Republic* must have been written later than *Gorgias* and before the *Laws*. These three works are longer than others, and take life within the *polis* largely for granted. The *polis* or city-state was the socio-political entity familiar to most Greeks, centring on an urban community but usually embracing enough productive land to satisfy the town's needs. Many were positioned near to the coast, facilitating maritime trade. The lengthy *Gorgias* had explored, primarily from the individual's point of view, the roles of justice and rhetoric within the *polis*. The *Republic* returns, at three times the length, to questions of justice and injustice, finding a parallel between its concept of justice within the state and justice within the individual. The *Laws*, allegedly left in an unfinished state at the time of Plato's death, is slightly longer still. It professes to legislate for the establishment and running of a Greek colony called 'Magnesia', often in great detail and often in ways that the casual reader finds of little appeal. It is more consistently 'political' and looks at the internal workings of the state largely from considerations of its own well-being, not that of its individual inhabitants. A fourth work, the rather late *Statesman* or *Politicus*, also has considerable political content, but the emphasis seems to be on the dialectical tools for analysing the statesman's role. It also seems to operate outside the confines of the Greek *polis*, using the term 'king' about as often as the term 'statesman', even though this term was largely used for foreign rulers of wide-ranging lands.

The Political and Intellectual Situation of Athens in Plato's Early Years

The Athens into which Plato was born had grown to be the largest Greek *polis* in terms of population and power, partly as a result of its extensive maritime trading links, partly through its silver resources, and partly through the wealth accumulated by the exploitation of its allies, which in a short time had turned them into what looked more like an empire. Some years before Plato's birth, the increasing tensions brought by Athenian success at the expense of others had brought about the Peloponnesian War, which lasted until 404 BCE, with a few years of uneasy peace (421–415 BCE). Athens' ultimate defeat saw it transformed from a successful democracy, in which all adult citizen males were expected to play an active role, into a period of political instability that led to the disastrous and oppressive regime of the thirty so-called 'tyrants' in 404–3 BCE, followed by a restored but weakened democracy. The instability had challenged the Athenian faith in what had shortly before been accepted political ideas. Democracy had been so transformed that the impulses of ordinary citizens, egged on by ambitious but unscrupulous public figures, could too easily lead the *polis* into disastrous decisions. Thinking Athenians consequently explored social and political ideas that they might never have contemplated if all had been running smoothly.

New ideas were also encouraged by a number of 'sophists', controversial intellectuals who roamed the Greek world searching for fee-paying pupils wanting to learn more of the many studies that had been gaining momentum in the eastern

Mediterranean. These subjects included rhetoric, grammar, cosmology, and everything from the pseudo-scientific study of religion and divination to new developments in medical theory. All these studies were undertaken principally in the hope of securing successful lives individually, usually within the *polis* but also in any afterlife that one might believe in.

Socrates and Socratic Discussions in Plato

One home-grown Athenian secured a reputation going far beyond that of most of these teachers, even though he did not demand a fee in return for his services. This was Socrates, who is credited with having turned philosophy away from its cosmological preoccupations and towards the study of ethics, by investigating above all else the nature and benefit of virtuous behaviour. He is noted for his interrogation of others on ethical matters, in particular for the exposure of inadequacies in their attempts to define ethical concepts. While Socrates was not himself an author, Socrates' followers, including Plato, wrote many dialogues that tried to capture the character of the dialectical exchanges between Socrates and others. We also have such works from the pen of Xenophon.

Socratic interrogation, usually known as *elenchus*, is conducted in a variety of early dialogues, such as *Charmides*, *Laches*, *Lysis*, and *Meno* (first part), but also in the first book of the *Republic*, which examines various attempts to explain or define what justice is. Socrates plays two roles here, that of narrator (for it is he who tells the story of the previous night's discussion), and that of chief speaker. As narrator he is able to comment on the action and the way in which the characters interacted as the conversation progressed, much as he had done in *Charmides* and *Lysis*. Thus, he controls the telling of the story. Definitions defended by the character Polemarchus, such as 'telling the truth and paying one's debts' and 'helping friends and harming enemies', soon have their weaknesses exposed, before the sophist Thrasymachus claims that justice is in fact 'what is in the interests of the stronger person'. In this case, it is not so much the definition that Socrates is anxious to refute, but rather the idea that injustice is more characteristic of knowledgeable action than justice. Thrasymachus is dragged reluctantly into admitting that this is not the case, but the discussion ends, as so often happens with Socrates' conversations, inconclusively. Those present are more inclined to admire Socrates' arguments than to believe their conclusions.

The Differences in Books II–X

The remaining books of the *Republic* continue similarly, to the extent that Socrates continues as narrator and as leading speaker. However, Socrates ceases to interrogate others and instead tries his best to satisfy their demands for him to explain his own views. The dynamic situation changes too, since the interested but less philosophical Polemarchus and the fierce and often sarcastic Thrasymachus withdraw into the background, and the conversation undergoes a new beginning initiated by Plato's elder brothers, Glaucon and Adeimantus, who are perceptive and polite, but firm. With a couple of notable exceptions, particularly at 449–450 where Polemarchus and Thrasymachus reappear as speakers, Socrates now ceases to comment on the action, and the narration adds very little apart from the addition of expressions like 'I said' and 'he said' from time to time. Hence, quite a lot of this reads more like the text of a drama than a narration. The majority of Plato's dialogues do in fact use a 'direct' or 'dramatic' presentation rather than the narrative one employed here, and it seems plausible that every dialogue begun after the *Republic*'s completion also chose dramatic form, and ceased to comment an any action except to the extent to which it could be embedded in the conversation itself.

Given the obvious differences between Book I and subsequent books, many scholars have been tempted to regard Book I as having started life as an independent dialogue of a more Socratic nature (e.g. G. Vlastos, *Socratic Studies*, 1994, 135; H. Thesleff, *Platonic Patterns*, 2009, 256–9, 429–30). It is rather similar in length to two other Socratic dialogues, narrated by Socrates, that consist mostly of unconvincing attempts to define ethical concepts, the *Charmides* and *Lysis*. Stylistic and linguistic differences from the rest of the *Republic* are not huge, but greater than those between other books of the dialogue. However, if Book I had been conceived as a separate work, then one would have to explain why some of its material, chiefly in the opening, is astonishingly appropriate as an introduction to the whole. The postulation of some sensitive adaptation once Book X had been written might account for this.

One reason for considering Book I separable is that when Book IX (588) goes back to the beginning of the argument, it goes back to the version of Thrasymachus' thesis discussed at the beginning of Book II, and does not associate that thesis directly with Thrasymachus. So perhaps Book II is the beginning of the argument proper, and in any case it is only here that one begins to meet the doctrines that we usually associate with the *Republic*. For these reasons, and certainly not because Book I is uninteresting or unattractive reading, it is not included in this abridged version.

Some Other Omissions

The other very large omission is the account of the degeneration of the allegedly best constitution ('Callipolis') in Books VIII-IX, which is not essential for anything that follows, and leaves no obvious mark either in the summary of the *Republic*'s constitution in Plato's later *Timaeus*, or in Aristotle's treatment of it in *Politics* II.1–5. In those books Plato sets out a theory of how his best state degenerates first into a 'timocracy' or warrior-state dominated by considerations of honour; then into oligarchy, where power resides with the wealthy and property is the primary motivator; then into democracy, which abhors such elitism but falls into worse mistakes when it gives way to all its whims and desires; then finally into tyranny, when a single champion of the people gathers resources to control the whole state single-handedly with nothing more in view than the one overriding desire that drives him on. All this seems unnecessary for understanding the arrangements that had earlier been made for the Platonic state itself. Indeed, a version of the *Republic* in six books rather than ten, known to us from citations in a single grammarian, might not have included this material.

This volume concentrates on Books II–V partly because it contains the core arrangements of Socrates' theoretical state – his *politeia* or republic from which the work takes its title; but also because it was probably *earlier* material (H. Thesleff, *Platonic Patterns*, 2009, 519–39), and essential for some of the material that follows. I talk as if each book was an obvious unit, but many of our book divisions are hardly natural breaks in the argument. The division between Book I and Book II seems well marked and entirely natural, matched to some extent by the divisions between Books VII and VIII. Book VIII begins with recapitulation that largely takes us back to Books II–V. To a lesser extent the breaks between IV and V and between IX and X also seem unobjectionable. Particularly difficult are the divisions between II and III and between VI and VII, which interrupt the discussion of poetry and the Sun–Line–Cave sequence respectively. We should thus regard the book divisions as partly due to practical considerations (e.g. how much text a single book-scroll could contain). The six-book version may also have involved divisions that were rather artificial.

Central to the *Republic* is the analogy between the *polis* and the human 'soul'. We have briefly discussed the *polis*, but something should be said about the soul, as that term has acquired various religious connotations in modern times. The word (Greek *psyche*) had come to be used for the core intellectual, sensory and emotional element of every human being. Other animals were often credited with a soul that was not so different. Whether such an entity was immortal or mortal, material or immaterial, unitary or complex, was a matter of much disagreement. Greeks often presumed that there was some kind of afterlife, but this was seldom seen as very desirable for the majority of people – dark and often quite insubstantial compared with life in this world. In the *Phaedo*, we meet the view usually associated with Plato's Socrates: that the soul was immaterial and immortal, that death was its separation from the body, and that its experiences thereafter would be determined by its conduct in this world until it could be born again.

The Concept of Soul

Differences in people's opinion about how a human being should behave were naturally dependent upon whether they believed in such an afterlife. If death really was the end for each person, then clear success or failure had to be measured purely on the basis of a single life. Surrendering any physical or material advantages would not be something lightly contemplated, and normally we should expect something in return, though not if such things were given up for the sake of one's children – for which reason our children were sometimes regarded as the only way in which we humans could expect some kind of limited future. Our children were the one thing we could profitably invest in for the long term. But if we ourselves can continue to exist as the same person beyond the grave, our success ought no longer to be measured in terms of this life, whether in respect of our material goods or our bodily health. The value of our children in offering us a shadowy future likewise diminishes. We should rather aim at the best possible future for our own soul, and our conduct within this life ought always to take the health of the soul into consideration.

Consequently, Socrates placed great emphasis on the care for our own souls. The soul would continue infinitely longer than the body, therefore its health was infinitely more important. The primary threat to it was perceived to be injustice or unfairness, so that just or fair conduct was always to be recommended. But how precisely was one to conceive of justice? And why was it such as would naturally have good consequences for us? How does injustice naturally have bad consequences? That, in a nutshell, is what *Republic* II–X sets out to determine, but the issues are complex.

From Justice in the State to Justice in the Individual

While the focus of Socratic discussions had ordinarily been on what virtuous and vicious behaviour meant for the individuals concerned, the *Republic* seeks to discover what justice would be in the virtuously run state in the hope that this will shed light on its meaning for the individual. Socrates sets about constructing and educating what he conceives of as a well-run state, and then searches in it for the four virtues that he expects to find there: wisdom, courage, temperance (sensibleness) and justice. He works on the principle that each person will have just one job, which means that besides specialist potters, cobblers, farmers and so on, there will be an army of specialist fighters (a novelty in the Greek *polis*) and specialist rulers: hence three distinct classes emerge. Each person is intended to the job for which he is best fitted, and for which he will be trained. Wisdom is expected to be found among the rulers, courage (a quality that preserves the correct opinion about what is to be feared and what is not) in the military, and temperance in the harmonious agreement that the rulers will govern and the others will submit to them. This leaves justice, which, somewhat strangely, is identified with whatever it is that ensures that everybody does the proper task.

Once justice in the state has been found, Socrates can turn to the individual, and asks whether the same divisions are to be found in the human soul. He argues that we have a reasoning part that ought to be in charge, like a ruler; a spirited part capable of anger and enthusiasm, and a desiring part that aims to satisfy the various desires of the body. He supports his division of soul by showing how two separate forces within us can simultaneously draw us in opposite directions. In the well-ordered individual, wisdom is found in the dominant rational part, courage in the spirited part, temperance in the harmonious agreement that reason is to be in charge, and justice in the state where each part persists with its own proper job without encroaching on the territory of other parts.

In both state and individual, then, there is a distinct plurality that must always remain discrete, and a unity that is always preserved by their complete cooperation. Both are one-and-many, as a mathematical subtext that sees goodness as unity seems often to be suggesting. Even in Book I, Socrates was trying to present justice as a unifying force, and injustice as a cause of dissension and division, so that a band of robbers would need an element of justice, allowing it to act cooperatively. It is not so difficult to see that the individuals least torn by competing internal forces and least at odds with others in their state are also likely to be happier than those torn by dissension, so that even in this life, promotion of justice both within one's self and in one's community will have beneficial results. So, Book IX claims that the just and kingly individual will be 729 times better off than the tyrannical person (the supreme example of injustice).

Republic as a Political Text: Poetry, the Sexes, and Children

With its analogy between the state and the individual, the *Republic* seems to switch with great ease between politics and ethics, so that it is not simply a contribution to political thought. Even so, it is studied more often as a political text, and there are good reasons for giving some weight to the principal political doctrines (or 'themes', if that is a better word). Among these are the rejection, for educational reasons, of the Greek poetic heritage, including the epic poetry of Homer and Hesiod and the dramas of the great Athenian tragedians. Both are seen as imitative or 'mimetic' because the poet, and subsequently the reader, imitates the kind of diction expected of the characters who speak, in most cases characters who fall far short of the intended ideal citizen. Even Platonism almost immediately found ways of denying this sweeping rejection by seeing much (or even all) of poetry as divinely inspired, and explaining away its allegedly corrupting passages as masking some perfectly innocent allegorical meaning. The rejection of such poetry, a huge part of Greek literature, is central to the educational program upon which the theoretical *polis* is built in Books II and III, and it returns again, backed with a new argument, in Book X.

Book V is where a number of controversial measures are defended, measures that seem to apply most directly to the superior class of 'guardians', as Aristotle observed in *Politics* II (1264a). They included the community of property, thought to eliminate the division between 'mine' and 'not-mine', thereby fostering more friendly relations between them; the community of spouses, which results from a tendency to see marriage purely as a device for the procreation of children, not from any consideration of what might be natural or satisfying; and the rejection of the notion that men and women, being different, should do entirely different jobs, as was widely the case in the ancient Greek world.

Rather, men and women are only different insofar as their procreational roles are concerned. Hence, there is no reason why only men should exercise in the gymnasia and go off to war. While men may be somewhat stronger and thus better equipped in some areas, there will be women who are actually better than most men at the pursuit that will determine their role in society. This enables them to become rulers of the state: 'guardianesses' who will thus be the superiors of most of the males in the *polis*. This should not be thought of as making Socrates and Plato proto-feminists, for in general the 'liberation' of their citizens is not among their priorities. Citizens endure restrictions that most of us would consider deeply intrusive, including strict controls over sex and reproduction that see the rulers deciding which possible couples would produce the best children, and then allocating sexual partners according to what is represented to the masses as a system of lots. Citizens are also lied to regularly by the rulers in these and other matters.

Nor will the children of the Guardians be brought up as the sons and daughters of this couple or that, but as the sons and daughters of all of them. Plato seems to feel that it is a good thing if sons do not know who their fathers are, because they would treat all males of the correct age (and class?) as their fathers in order to avoid the terrible crime of patricide. But spare a thought for that embryo or child that is conceived contrary to the laws of the state, or with some deformity. The ideal state finds no place for them.

Philosopher-Rulers and Higher Education

One wonders whether Plato, in the light of his later disastrous intervention in Syracusan politics, still continued to imagine that it should be the philosophers, rather than any less outstanding Guardians, sometimes known as auxiliaries, who rule. The discussion of the philosophers and their claim to rule continues far into Book VI. This claim is based largely on the fact that philosophers (literally 'wisdom-lovers') stand opposed to 'sight-lovers', and are primarily concerned with observing true reality: what each thing is in itself, often called Platonic 'Ideas' or 'Forms'. Indeed, others are content with mere opinion, for that is all the unstable realities of the world of sensation can yield. Stable reality is sought through dialectical argument, and will include such objects as the Justice and Courage that Socrates has defined earlier. And at the pinnacle of this world, to which philosophy alone can lead us, is the Idea of the Good – the highest object of learning, as Book VI (504) calls it. Clearly we should like our politicians to have a clear idea of some genuine good at which to aim – some good that can never be anything but good. So the belief that only the philosopher can discern such realities leads inevitably to the episode that now runs through to the end of Book VII, dealing with the education of the philosophers themselves: their education for rule, and the requirement that they should then devote themselves to repaying the state by ruling well.

Here we meet the comparison between the sun of the physical world and the Good of the intelligible world, casting a 'light' upon the entire intelligible world and nourishing it. This may be intended as the pinnacle of the whole *Republic.* We now meet a geometrical figure, the Divided Line, that will demonstrate the various images of original and image that are found within and between the intelligible and sensible worlds. And then, at the beginning of Book VII, the image of the Cave, that presents in a more pictorial fashion all these relationships. We are like prisoners, able to view only the shadows of artificial objects projected onto the back wall of a cave. The result of release from this condition, probably seen as analogous to that of the spectator in the Athenian theatre, is a journey to see the objects themselves, then the shadows and reflections outside the cave, the outside world more generally, and finally the very sun itself that illuminates the outside world. Additional education will secure this turning upwards: the mathematical sciences and, beyond this, dialectic.

The Concluding Books

There follow Books VIII and IX, mostly concerned with inferior political constitutions and inferior states of soul, culminating in the supreme picture of the wretched individual: the tyrant. The contrast between the ideal philosopher-king and the tyrant finally shows the rewards in this life that justice brings and injustice denies. This is emphasized in the image of the many-headed beast (within us), which, like the lion within, must learn to obey the internal human, i.e. the reason. The tripartite soul from Book IV has resurfaced in a slightly different form. The end of Book IX serves as a this-worldly conclusion. There remains Book X, which subjects poetry to a further critique and explains its failings, this time in terms of its distorting effect on the souls of its audience. For the soul can allegedly be shown to be immortal, and damaging it is a longer-lasting blow than damaging the body. If the argument, based on the inability of the soul's greatest evil (injustice!) to destroy it, should fail to convince, Plato adds one of his finest myths, the Myth of Er, that tells of the fate of souls in the afterlife, and of the superior position of the educated soul that can carefully choose what life it will have next, in what is perhaps the most dangerous moment of all.

This Translation and Abridgement

The translation of Jowett used here revives the work of one of the most respected Platonists of Victorian times, who resisted attempts to turn Plato into a mere system (M. Schofield, *How Plato Writes*, 2023, 52–67). The numbers included from time to time, in square brackets, match those widely employed in Greek texts and translations alike, following the early printed edition of Stephanus. They are placed at the start of sentences in which the original page-break occurred. Omissions have been indicated. In Books II–V these are generally minor, less than one Stephanus page, and often involve further examples or material not easily understood without much background knowledge. An occasional unnecessary word of explanation added by Jowett has been replaced only by an ellipsis. Larger omissions in Books

VI–X have been explained more fully, and what remains is an attempt to offer the reader as full an appreciation as possible of the entire integrated work.

Harold Tarrant

Plato

The Republic

Translated by Benjamin Jowett

Abridged by Harold Tarrant

Persons of the Dialogue

Socrates, who is the narrator
Glaucon
Adeimantus
Polemarchus
Thrasymachus
And others who are auditors, including Cephalus and Cleitophon who speak only in Book I

The scene is laid in the house of Cephalus at the Piraeus; and the whole dialogue is narrated by Socrates the day after it actually took place.

Book I has been omitted for reasons given in the introduction. Stephanus number references are marked in square brackets.

BOOK II

[357]

With these words I was thinking that I had made an end of the discussion; but the end, in truth, proved to be only a beginning. For Glaucon, who is always the most pugnacious of men, was dissatisfied at Thrasymachus' retirement; he wanted to have the battle out. So he said to me: Socrates, do you wish really to persuade us, or only to seem to have persuaded us, that to be just is always better than to be unjust?

I should wish really to persuade you, I replied, if I could.

Then you certainly have not succeeded. Let me ask you now: – How would you arrange goods – are there not some which we welcome for their own sakes, and independently of their consequences, as, for example, harmless pleasures and enjoyments, which delight us at the time, although nothing follows from them?

I agree in thinking that there is such a class, I replied.

Is there not also a second class of goods, such as knowledge, sight, health, which are desirable not only in themselves, but also for their results?

Certainly, I said.

And would you not recognize a third class, such as gymnastic, and the care of the sick, and the physician's art; also the various ways of money-making – these do us good but we regard them as disagreeable; and no one would choose them for their own sakes, but only for the sake of some reward or result which flows from them?

There is, I said, this third class also. But why do you ask?

Because I want to know in which of the three classes you would place justice?

[358] In the highest class, I replied, – among those goods which he who would be happy desires both for their own sake and for the sake of their results.

Then the many are of another mind; they think that justice is to be reckoned in the troublesome class, among goods which are to be pursued for the sake of rewards and of reputation, but in themselves are disagreeable and rather to be avoided.

I know, I said, that this is their manner of thinking, and that this was the thesis which Thrasymachus was maintaining just now, when he censured justice and praised injustice. But I am too stupid to be convinced by him.

I wish, he said, that you would hear me as well as him, and then I shall see whether you and I agree. For Thrasymachus seems to me, like a snake, to have been charmed by your voice sooner than he ought to have been; but to my mind the nature of justice and injustice have not yet been made clear. Setting aside their rewards and results, I want to know what they are in themselves, and how they inwardly work in the soul. If you, please, then, I will revive the argument of Thrasymachus. And first I will speak of the nature and origin of justice according to the common view of them. Secondly, I will show that all men who practise justice do so against their will, of necessity, but not as a good. And thirdly, I will argue that there is reason in this view, for the life of the unjust is after all better far than the life of the just – if what they say is true, Socrates, since I myself am not of their opinion. But still I acknowledge that I am perplexed when I hear the voices of Thrasymachus and myriads of others dinning in my ears; and, on the other hand, I have never yet heard the superiority of justice to injustice maintained by anyone in a satisfactory way. I want to hear justice praised in respect of itself; then I shall be satisfied, and you are the person from whom I think that I am most likely to hear this; and therefore I will praise the unjust life to the utmost of my power, and my manner of speaking will indicate the manner in which I desire to hear you too praising justice and censuring injustice. Will you say whether you approve of my proposal?

Indeed I do; nor can I imagine any theme about which a man of sense would oftener wish to converse.

I am delighted, he replied, to hear you say so, and shall begin by speaking, as I proposed, of the nature and origin of justice.

They say that to do injustice is, by nature, good; to suffer injustice, evil; but that the evil is greater than the good. And so when men have both done and suffered injustice and have had experience of both, not being able to [359] avoid the one and obtain the other, they think that they had better agree among themselves to have neither; hence there arise laws and mutual covenants; and that which is ordained by law is termed by them lawful and just. This they affirm to be the origin and nature of justice; – it is a mean or compromise, between the best of all, which is to do injustice and not be punished, and the worst of all, which is to suffer injustice without the power of retaliation; and justice, being at a middle point between the two, is tolerated not as a good, but as the lesser evil, and honoured by reason of the inability of men to do injustice. For no man who is worthy to be called a man would ever submit to such an agreement if he were able to resist; he would be mad if he did. Such is the received account, Socrates, of the nature and origin of justice.

Now that those who practise justice do so involuntarily and because they have not the power to be unjust will best appear if we imagine something of this kind: having given both to the just and the unjust power to do what they will, let us watch and see whither desire will lead them; then we shall discover in the very act the just and unjust man to be proceeding along the same road, following their interest, which all natures deem to be their good, and are only diverted into the path of justice by the force of law. The liberty which we are supposing may be most completely given to them in the form of such a power as is said to have been possessed by Gyges, the ancestor of Croesus the Lydian. According to the tradition, Gyges was a shepherd in the service of the king of Lydia; there was a great storm, and an earthquake made an opening in the earth at the place where he was feeding his flock. Amazed at the sight, he descended into the opening, where, among other marvels, he beheld a hollow brazen horse, having doors, at which he stooping and looking in saw a dead body of stature, as appeared to him, more than human, and having nothing on but a gold ring; this he took from the finger of the dead and reascended. Now the shepherds met together, according to custom, that they might send their monthly report about the flocks to the king; into their assembly he came having the ring on his finger, and as he was sitting among them he chanced to turn the collet of the ring inside his hand, [360] when instantly he became invisible to the rest of the company and they began to speak of him as if he were no longer present. He was astonished at this, and again touching the ring he turned the collet outwards and reappeared; he made several trials of the ring, and always with the same result – when he turned the collet inwards he became invisible, when outwards he reappeared. Whereupon he contrived to be chosen one of the messengers who were sent to the court; whereas soon as he arrived he seduced the queen, and with her help conspired against the king and slew him, and took the kingdom. Suppose now that there were two such magic rings, and the just put on one of them and the unjust the other; no man can be imagined to be of such an iron nature that he would stand fast in justice. No man would keep his hands off what was not his own when he could safely take what he liked out of the market, or go into houses and lie with any one at his pleasure, or kill or release from prison whom he would, and in all respects be like a God among men. Then the actions of the just would be as the actions of the unjust; they would both come at last to the same point. And this we may truly affirm to be a great proof that a man is just, not willingly or because he thinks that justice is any good to him individually, but of necessity, for wherever anyone thinks that he can safely be unjust, there he is unjust. For all men believe in their hearts that injustice is far more profitable to the individual than justice, and he who argues as I have been supposing, will say that they are right. If you could imagine anyone obtaining this power of becoming invisible, and never doing any wrong or touching what was another's, he would be thought by the lookers-on to be a most wretched idiot, although they would praise him to one another's faces, and keep up appearances with one another from a fear that they too might suffer injustice. Enough of this.

Now, if we are to form a real judgment of the life of the just and unjust, we must isolate them; there is no other way; and how is the isolation to be effected? I answer: Let the unjust man be entirely unjust, and the just man entirely just; nothing is to be taken away from either of them, and both are to be perfectly furnished for the work of their respective lives. First, let the unjust be like other distinguished masters of craft; like the skilful pilot or physician, who knows intuitively his own powers [361] and keeps within their limits, and who, if he fails at any point, is able to recover himself. So let the unjust make his unjust attempts in the right way, and lie hidden if he means to be great in his injustice: (he who is found out is nobody) for the highest reach of injustice is to be deemed just when you are not. Therefore, I say that in the perfectly unjust man we must assume the most perfect injustice; there is to be no deduction, but we must allow him, while doing the most unjust acts, to have acquired the greatest reputation for justice. If he have taken a false step, he must be able to recover himself; he must be one who can speak with effect, if any of his deeds come to light, and who can force his way where force is required by his courage and strength, and command of money and friends. And at his side let us place the just man in his nobleness and simplicity, wishing, as Aeschylus says, to be and not to seem good. There must be no seeming, for if he seem to be just he will be honoured and rewarded, and then we shall not know whether he is just for the sake of justice or for the sake of honours and rewards; therefore, let him be clothed in justice only, and have no other covering; and he must be imagined in a state of life the opposite of the former. Let him be the best of men, and let him be thought the worst; then he will have been put to the proof; and we shall see whether he will be affected by the fear of infamy and its consequences. And let him continue thus to the hour of death; being just and seeming to be unjust. When both have reached the uttermost extreme, the one of justice and the other of injustice, let judgment be given which of them is the happier of the two.

Heavens! my dear Glaucon, I said, how energetically you polish them up for the decision, first one and then the other, as if they were two statues.

I do my best, he said. And now that we know what they are like there is no difficulty in tracing out the sort of life which awaits either of them. This I will proceed to describe; but as you may think the description a little too coarse, I ask you to suppose, Socrates, that the words which follow are not mine. – Let me put them into the mouths of the eulogists of injustice: They will tell you that the just man who is thought unjust will be scourged, racked, bound – will have his eyes burnt out; [362] and, at last, after suffering every kind of evil, he will be impaled: Then he will understand that he ought to seem only, and not to be, just; the words of Aeschylus may be more truly spoken of the unjust than of the just. For the unjust is pursuing a reality; he does not live with a view to appearances – he wants to be really unjust and not to seem only: –

'His mind has a soil deep and fertile, out of which spring his prudent counsels.'

In the first place, he is thought just, and therefore bears rule in the city; he can marry whom he will, and give in marriage to whom he will; also he can trade and deal where he likes, and always to his own advantage, because he has no misgivings about injustice; and at every contest, whether in public or private, he gets the better of his antagonists, and gains at their expense, and is rich, and out of his gains he can benefit his friends, and harm his enemies; moreover, he can offer sacrifices, and dedicate gifts to the gods abundantly and magnificently, and can honour the gods or any man whom he wants to honour in a far better style than the just, and therefore he is likely to be dearer than they are to the gods. And thus, Socrates, gods and men are said to unite in making the life of the unjust better than the life of the just.

I was going to say something in answer to Glaucon, when Adeimantus, his brother, interposed: Socrates, he said, you do not suppose that there is nothing more to be urged?

Why, what else is there? I answered.

The strongest point of all has not been even mentioned, he replied.

Well, then, according to the proverb, 'Let brother help brother' – if he fails in any part do you assist him; although I must confess that Glaucon has already said quite enough to lay me in the dust, and take from me the power of helping justice.

Nonsense, he replied. But let me add something more: There is another side to Glaucon's argument about the praise and censure of justice and injustice, which is equally required in order to bring out what I believe to be his meaning. Parents and tutors are always telling their sons [363] and their wards that they are to be just; but why? not for the sake of justice, but for the sake of character and reputation; in the hope of obtaining for him who is reputed just some of those offices, marriages, and the like which Glaucon has enumerated among the advantages accruing to the unjust from the reputation of justice. More, however, is made of appearances by this class of persons than by the others; for they throw in the good opinion of the gods, and will tell you of a shower of benefits which the heavens, as they say, rain upon the pious; and this accords with the testimony of the noble Hesiod and Homer, the first of whom says, that the gods make the oaks of the just –

'To bear acorns at their summit, and bees in the middle;

and the sheep are bowed down with the weight of their fleeces,'

And many other blessings of a like kind are provided for them. And Homer has a very similar strain; for he speaks of one whose fame is –

'As the fame of some blameless king who, like a god, maintains justice; to whom the black earth brings forth wheat and barley, whose trees are bowed with fruit, and his sheep never fail to bear, and the sea gives him fish.'

Still grander are the gifts of heaven which Musaeus and his son vouchsafe to the just; they take them down into the world below, where they have the saints lying on couches at a feast, everlastingly drunk, crowned with garlands; their idea seems to be that an immortality of drunkenness is the highest meed of virtue. Some extend their rewards yet further; the posterity, as they say, of the faithful and just shall survive to the third and fourth generation. This is the style in which they praise justice. But about the wicked there is another strain; they bury them in a slough in Hades, and make them carry water in a sieve; also while they are yet living they bring them to infamy, and inflict upon them the punishments which Glaucon described as the portion of the just who are reputed to be unjust; nothing else does their invention supply. Such is their manner of praising the one and censuring the other.

Once more, Socrates, I will ask you to consider another way of speaking about justice and injustice, which is not confined to the poets, but is found in prose writers. [364] The universal voice of mankind is always declaring that justice and virtue are honourable, but grievous and toilsome; and that the pleasures of vice and injustice are easy of attainment, and are only censured by law and opinion. They say also that honesty is for the most part less profitable than dishonesty; and they are quite ready to call wicked men happy, and to honour them both in public and private when they are rich or in any other way influential, while they despise and overlook those who may be weak and poor, even though acknowledging them to be better than the others. But most extraordinary of all is their mode of speaking about virtue and the gods: they say that the gods apportion calamity and misery to many good men, and good and happiness to the wicked. And mendicant prophets go to rich men's doors and persuade them that they have a power committed to them by the gods of making an atonement for a man's own or his ancestor's sins by sacrifices or charms, with rejoicings and feasts; and they promise to harm an enemy, whether just or unjust, at a small cost; with magic arts and incantations binding heaven, as they say, to execute their will. And the poets are the authorities to whom they appeal, now smoothing the path of vice with the words of Hesiod: –

'Vice may be had in abundance without trouble; the way is smooth and her dwelling-place is near. But before virtue the gods have set toil,'

And a tedious and uphill road: then citing Homer as a witness that the gods may be influenced by men; for he also says: –

'The gods, too, may be turned from their purpose; and men pray to them and avert their wrath by sacrifices and soothing entreaties, and by libations and the odour of fat, when they have sinned and transgressed.'

And they produce a host of books written by Musaeus and Orpheus, who were children of the Moon and the Muses – that is what they say – according to which they perform their ritual, and persuade not only individuals, but whole cities, that expiations and atonements for sin may be made by sacrifices and amusements which fill a vacant hour, [365] and are equally at the service of the living and the dead; the latter sort they call mysteries, and they redeem us from the pains of hell, but if we neglect them no one knows what awaits us.

He proceeded: And now when the young hear all this said about virtue and vice, and the way in which gods and men regard them, how are their minds likely to be affected, my dear Socrates, – those of them, I mean, who are quickwitted, and, like bees on the wing, light on every flower, and from all that they hear are prone to draw conclusions as to what manner of persons they should be and in what way they should walk if they would make the best of life? Probably the youth will say to himself in the words of Pindar –

'Can I by justice or by crooked ways of deceit ascend a loftier tower which may be a fortress to me all my days?'

For what men say is that, if I am really just and am not also thought just profit there is none, but the pain and loss on the other hand are unmistakeable. But if, though unjust, I acquire the reputation of justice, a heavenly life is promised to me. Since then, as philosophers prove, appearance tyrannizes over truth and is lord of happiness, to appearance I must devote myself. I will describe around me a picture and shadow of virtue to be the vestibule and exterior of my house; behind I will trail the subtle and crafty fox, as Archilochus, greatest of sages, recommends. But I hear someone exclaiming that the concealment of wickedness is often difficult; to which I answer, Nothing great is easy. Nevertheless, the argument indicates this, if we would be happy, to be the path along which we should proceed. With a view to concealment we will establish secret brotherhoods and political clubs. And there are professors of rhetoric who teach the art of persuading courts and assemblies; and so, partly by persuasion and partly by force, I shall make unlawful gains and not be punished. Still I hear a voice saying that the gods cannot be deceived, neither can they be compelled. But what if there are no gods? Or, suppose them to have no care of human things – why in either case should we mind about concealment? And even if there are gods, and they do care about us, yet we know of them only from tradition and the genealogies of the poets; and these are the very persons who say that they may be influenced and turned by 'sacrifices and soothing entreaties and by offerings.' Let us be consistent then, and believe both or neither. If the poets speak truly, why then we had better be unjust, [366] and offer of the fruits of injustice; for if we are just, although we may escape the vengeance of heaven, we shall lose the gains of injustice; but, if we are unjust, we shall keep the gains, and by our sinning and praying, and praying and sinning, the gods will be propitiated, and we shall not be punished. 'But there is a world below in which either we or our posterity will suffer for our unjust deeds.' Yes, my friend, will be the reflection, but there are mysteries and atoning deities, and these have great power. That is what mighty cities declare; and the children of the gods, who were their poets and prophets, bear a like testimony.

On what principle, then, shall we any longer choose justice rather than the worst injustice? when, if we only unite the latter with a deceitful regard to appearances, we shall fare to our mind both with gods and men, in life and after death, as the most numerous and the highest authorities tell us. Knowing all this, Socrates, how can a man who has any superiority of mind or person or rank or wealth, be willing to honour justice; or indeed to refrain from laughing when he hears justice praised? And even if there should be some one who is able to disprove the truth of my words, and who is satisfied that justice is best, still he is not angry with the unjust, but is very ready to forgive them, because he also knows that men are not just of their own free will; unless, peradventure, there be some one whom the divinity within him may have inspired with a hatred of injustice, or who has attained knowledge of the truth – but no other man. He only blames injustice who, owing to cowardice or age or some weakness, has not the power of being unjust. And this is proved by the fact that when he obtains the power, he immediately becomes unjust as far as he can be.

The cause of all this, Socrates, was indicated by us at the beginning of the argument, when my brother and I told you how astonished we were to find that of all the professing panegyrists of justice – beginning with the ancient heroes of whom any memorial has been preserved to us, and ending with the men of our own time – no one has ever blamed injustice or praised justice except with a view to the glories, honours, and benefits which flow from them. No one has ever adequately described either in verse or prose the true essential nature of either of them abiding in the soul, and invisible to any human or divine eye; or shown that of all the things of a man's soul which he has within him, justice is the greatest good, [367] and injustice the greatest evil. Had this been the universal strain, had you sought to persuade us of this from our youth upwards, we should not have been on the watch to keep one another from doing wrong, but everyone would have been his own watchman, because afraid, if he did wrong, of harbouring in himself the greatest of evils. I dare say that Thrasymachus and others would seriously hold the language which I have been merely repeating, and words even stronger than these about justice and injustice, grossly, as I conceive, perverting their true nature. But I speak in this vehement manner, as I must frankly confess to you, because I want to hear from you the opposite side; and I would ask you to show not only the superiority which justice has over injustice, but what effect they have on the possessor of them which makes the one to be a good and the other an evil to him. And please, as Glaucon requested of you, to exclude reputations; for unless you take away from each of them his true reputation and add on the false, we shall say that you do not praise justice, but the appearance of it; we shall think that you are only exhorting us to keep injustice dark, and that you really agree with Thrasymachus in thinking that justice is another's good and the interest of the stronger, and that injustice is a man's own profit and interest, though injurious to the weaker. Now, as you have admitted that justice is one of that highest class of goods which are desired indeed for their results, but in a far greater degree for their own sakes – like sight or hearing or knowledge or health, or any other real and natural and not merely conventional good – I would ask you in your praise of justice to regard one point only: I mean the essential good and evil which justice and injustice work in the possessors of them. Let others praise justice and censure injustice, magnifying the rewards and honours of the one and abusing the other; that is a manner of arguing which, coming from them, I am ready to tolerate, but from you who have spent your whole life in the consideration of this question, unless I hear the contrary from your own lips, I expect something better. And therefore, I say, not only prove to us that justice is better than injustice, but show what they either of them do to the possessor of them, which makes the one to be a good and the other an evil, whether seen or unseen by gods and men.

I had always admired the genius of Glaucon and Adeimantus, but on hearing these words I was quite delighted,

[368] and said: Sons of an illustrious father, that was not a bad beginning of the Elegiac verses which the admirer of Glaucon made in honour of you after you had distinguished yourselves at the battle of Megara: –

'Sons of Ariston,' he sang, 'divine offspring of an illustrious hero.'

The epithet is very appropriate, for there is something truly divine in being able to argue as you have done for the superiority of injustice, and remaining unconvinced by your own arguments. And I do believe that you are not convinced – this I infer from your general character, for had I judged only from your speeches I should have mistrusted you. But now, the greater my confidence in you, the greater is my difficulty in knowing what to say. For I am in a strait between two; on the one hand I feel that I am unequal to the task; and my inability is brought home to me by the fact that you were not satisfied with the answer which I made to Thrasymachus, proving, as I thought, the superiority which justice has over injustice. And yet I cannot refuse to help, while breath and speech remain to me; I am afraid that there would be an impiety in being present when justice is evil spoken of and not lifting up a hand in her defence. And therefore I had best give such help as I can.

Glaucon and the rest entreated me by all means not to let the question drop, but to proceed in the investigation. They wanted to arrive at the truth, first, about the nature of justice and injustice, and secondly, about their relative advantages. I told them, what I really thought, that the enquiry would be of a serious nature, and would require very good eyes. Seeing then,

I said, that we are no great wits, I think that we had better adopt a method which I may illustrate thus: suppose that a short-sighted person had been asked by someone to read small letters from a distance; and it occurred to someone else that they might be found in another place which was larger and in which the letters were larger – if they were the same and he could read the larger letters first, and then proceed to the lesser – this would have been thought a rare piece of good fortune.

Very true, said Adeimantus; but how does the illustration apply to our enquiry?

I will tell you, I replied; justice, which is the subject of our enquiry, is, as you know, sometimes spoken of as the virtue of an individual, and sometimes as the virtue of a State.

True, he replied.

And is not a State larger than an individual?

It is.

Then in the larger the quantity of justice is likely to be larger and [369] more easily discernible. I propose therefore that we enquire into the nature of justice and injustice, first as they appear in the State, and secondly in the individual, proceeding from the greater to the lesser and comparing them.

That, he said, is an excellent proposal.

And if we imagine the State in process of creation, we shall see the justice and injustice of the State in process of creation also.

I dare say.

When the State is completed there may be a hope that the object of our search will be more easily discovered.

Yes, far more easily.

But ought we to attempt to construct one? I said; for to do so, as I am inclined to think, will be a very serious task. Reflect therefore.

I have reflected, said Adeimantus, and am anxious that you should proceed.

A State, I said, arises, as I conceive, out of the needs of mankind; no one is self-sufficing, but all of us have many wants. Can any other origin of a State be imagined?

There can be no other.

Then, as we have many wants, and many persons are needed to supply them, one takes a helper for one purpose and another for another; and when these partners and helpers are gathered together in one habitation, the body of inhabitants is termed a State.

True, he said.

And they exchange with one another, and one gives, and another receives, under the idea that the exchange will be for their good.

Very true.

Then, I said, let us begin and create in idea a State; and yet the true creator is necessity, who is the mother of our invention.

Of course, he replied.

Now the first and greatest of necessities is food, which is the condition of life and existence.

Certainly.

The second is a dwelling, and the third clothing and the like.

True.

And now let us see how our city will be able to supply this great demand: We may suppose that one man is a husbandman, another a builder, someone else a weaver – shall we add to them a shoemaker, or perhaps some other purveyor to our bodily wants?

Quite right.

The barest notion of a State must include four or five men.

Clearly.

And how will they proceed? Will each bring the result of his labours into a common stock? – the individual husbandman, for example, producing for four, and labouring four times as long and as much as he need in the provision of food with which he supplies others as well as himself; or will he have nothing to do with others and not be at the trouble of producing for them, but provide for himself alone a fourth of the [370] food in a fourth of the time, and in the remaining three fourths of his time be employed in making a house or a coat or a pair of shoes, having no partnership with others, but supplying himself all his own wants?

Adeimantus thought that he should aim at producing food only and not at producing everything.

Probably, I replied, that would be the better way; and when I hear you say this, I am myself reminded that we are not all alike; there are diversities of natures among us which are adapted to different occupations.

Very true.

And will you have a work better done when the workman has many occupations, or when he has only one?

When he has only one.

Further, there can be no doubt that a work is spoilt when not done at the right time?

No doubt.

For business is not disposed to wait until the doer of the business is at leisure; but the doer must follow up what he is doing, and make the business his first object.

He must.

And if so, we must infer that all things are produced more plentifully and easily and of a better quality when one man does one thing which is natural to him and does it at the right time, and leaves other things.

Undoubtedly.

Then more than four citizens will be required; for the husbandman will not make his own plough or mattock, or other implements of agriculture, if they are to be good for anything. Neither will the builder make his tools – and he too needs many; and in like manner the weaver and shoemaker.

True.

Then carpenters, and smiths, and many other artisans, will be sharers in our little State, which is already beginning to grow?

True.

Yet even if we add neatherds, shepherds, and other herdsmen, in order that our husbandmen may have oxen to plough with, and builders as well as husbandmen may have draught cattle, and curriers and weavers fleeces and hides, – still our State will not be very large.

That is true; yet neither will it be a very small State which contains all these.

Then, again, there is the situation of the city – to find a place where nothing need be imported is wellnigh impossible.

Impossible.

Then there must be another class of citizens who will bring the required supply from another city?

There must.

But if the trader goes empty-handed, having nothing which they [371] require who would supply his need, he will come back empty-handed.

That is certain.

And therefore what they produce at home must be not only enough for themselves, but such both in quantity and quality as to accommodate those from whom their wants are supplied.

Very true.

Then more husbandmen and more artisans will be required?

They will.

Not to mention the importers and exporters, who are called merchants?

Yes.

Then we shall want merchants?

We shall.

And if merchandise is to be carried over the sea, skilful sailors will also be needed, and in considerable numbers?

Yes, in considerable numbers.

Then, again, within the city, how will they exchange their productions? To secure such an exchange was, as you will remember, one of our principal objects when we formed them into a society and constituted a State.

Clearly they will buy and sell.

Then they will need a market-place, and a money-token for purposes of exchange.

Certainly.

Suppose now that a husbandman, or an artisan, brings some production to market, and he comes at a time when there is no one to exchange with him, – is he to leave his calling and sit idle in the market-place?

Not at all; he will find people there who, seeing the want, undertake the office of salesmen. In well-ordered states they are commonly those who are the weakest in bodily strength, and therefore of little use for any other purpose; their duty is to be in the market, and to give money in exchange for goods to those who desire to sell and to take money from those who desire to buy.

This want, then, creates a class of retail-traders in our State. Is not 'retailer' the term which is applied to those who sit in the market-place engaged in buying and selling, while those who wander from one city to another are called merchants?

Yes, he said.

And there is another class of servants, who are intellectually hardly on the level of companionship; still they have plenty of bodily strength for labour, which accordingly they sell, and are called, if I do not mistake, hirelings, hire being the name which is given to the price of their labour.

True.

Then hirelings will help to make up our population?

Yes.

And now, Adeimantus, is our State matured and perfected?

I think so.

Where, then, is justice, and where is injustice, and in what part of the State did they spring up?

[372] Probably in the dealings of these citizens with one another. I cannot imagine that they are more likely to be found anywhere else.

I dare say that you are right in your suggestion, I said; we had better think the matter out, and not shrink from the enquiry.

Let us then consider, first of all, what will be their way of life, now that we have thus established them. Will they not produce corn, and wine, and clothes, and shoes, and build houses for themselves? And when they are housed, they will work, in summer, commonly, stripped and barefoot, but in winter substantially clothed and shod. They will feed on barley-meal and flour of wheat, baking and kneading them, making noble cakes and loaves; these they will serve up on a mat of reeds or on clean leaves, themselves reclining the while upon beds strewn with yew or myrtle. And they and their children will feast, drinking of the wine which they have made, wearing garlands on their heads, and hymning the praises of the gods, in happy converse with one another. And they will take care that their families do not exceed their means; having an eye to poverty or war.

But, said Glaucon, interposing, you have not given them a relish to their meal.

True, I replied, I had forgotten; of course they must have a relish – salt, and olives, and cheese, and they will boil roots and herbs such as country people prepare; for a dessert we shall give them figs, and peas, and beans; and they will roast myrtle-berries and acorns at the fire, drinking in moderation. And with such a diet they may be expected to live in peace and health to a good old age, and bequeath a similar life to their children after them.

Yes, Socrates, he said, and if you were providing for a city of pigs, how else would you feed the beasts?

But what would you have, Glaucon? I replied.

Why, he said, you should give them the ordinary conveniences of life. People who are to be comfortable are accustomed to lie on sofas, and dine off tables, and they should have sauces and sweets in the modern style.

Yes, I said, now I understand: the question which you would have me consider is, not only how a State, but how a luxurious State is created; and possibly there is no harm in this, for in such a State we shall be more likely to see how justice and injustice originate. In my opinion the true and healthy constitution of the State is the one which I have described. But if you wish also to see a State at fever-heat, I have no objection. [373] For I suspect that many will not be satisfied with the simpler way of life. They will be for adding sofas, and tables, and other furniture; also dainties, and perfumes, and incense, and courtesans, and cakes, all these not of one sort only, but in every variety; we must go beyond the necessaries of which I was at first speaking, such as houses, and clothes, and shoes: the arts of the painter and the embroiderer will have to be set in motion, and gold and ivory and all sorts of materials must be procured.

True, he said.

Then we must enlarge our borders; for the original healthy State is no longer sufficient. Now will the city have to fill and swell with a multitude of callings which are not required by any natural want; such as the whole tribe of hunters and actors, of whom one large class have to do with forms and colours; another will be the votaries of music – poets and their attendant train of rhapsodists, players, dancers, contractors; also makers of divers kinds of articles, including women's dresses. And we shall want more servants. Will not tutors be also in request, and nurses wet and dry, tirewomen and barbers, as well as confectioners and cooks; and swineherds, too, who were not needed and therefore had no place in the former edition of our State, but are needed now? They must not be forgotten: and there will be animals of many other kinds, if people eat them.

Certainly.

And living in this way we shall have much greater need of physicians than before?

Much greater.

And the country which was enough to support the original inhabitants will be too small now, and not enough?

Quite true.

Then a slice of our neighbours' land will be wanted by us for pasture and tillage, and they will want a slice of ours, if, like ourselves, they exceed the limit of necessity, and give themselves up to the unlimited accumulation of wealth?

That, Socrates, will be inevitable.

And so we shall go to war, Glaucon. Shall we not?

Most certainly, he replied.

Then without determining as yet whether war does good or harm, thus much we may affirm, that now we have discovered war to be derived from causes which are also the causes of almost all the evils in States, private as well as public.

Undoubtedly.

And our State must once more enlarge; and this time the enlargement will be nothing short [374] of a whole army, which will have to go out and fight with the invaders for all that we have, as well as for the things and persons whom we were describing above.

Why? he said; are they not capable of defending themselves?

No, I said; not if we were right in the principle which was acknowledged by all of us when we were framing the State: the principle, as you will remember, was that one man cannot practise many arts with success.

Very true, he said.

But is not war an art?

Certainly.

And an art requiring as much attention as shoemaking?

Quite true.

And the shoemaker was not allowed by us to be a husbandman, or a weaver, or a builder – in order that we might have our shoes well made; but to him and to every other worker was assigned one work for which he was by nature fitted, and at that he was to continue working all his life long and at no other; he was not to let opportunities slip, and then he would become a good workman. Now nothing can be more important than that the work of a soldier should be well done. But is war an art so easily acquired that a man may be a warrior who is also a husbandman, or shoemaker, or other artisan; although no one in the world would be a good dice or draught player who merely took up the game as a recreation, and had not from his earliest years devoted himself to this and nothing else? No tools will make a man a skilled workman, or master of defence, nor be of any use to him who has not learned how to handle them, and has never bestowed any attention upon them. How then will he who takes up a shield or other implement of war become a good fighter all in a day, whether with heavy-armed or any other kind of troops?

Yes, he said, the tools which would teach men their own use would be beyond price.

And the higher the duties of the guardian, I said, the more time, and skill, and art, and application will be needed by him?

No doubt, he replied.

Will he not also require natural aptitude for his calling?

Certainly.

Then it will be our duty to select, if we can, natures which are fitted for the task of guarding the city?

It will.

And the selection will be no easy matter, I said; but we must be brave and do our best.

[375] We must.

Is not the noble youth very like a well-bred dog in respect of guarding and watching?

What do you mean?

I mean that both of them ought to be quick to see, and swift to overtake the enemy when they see him; and strong too if, when they have caught him, they have to fight with him.

All these qualities, he replied, will certainly be required by them.

Well, and your guardian must be brave if he is to fight well?

Certainly.

And is he likely to be brave who has no spirit, whether horse or dog or any other animal? Have you never observed how invincible and unconquerable is spirit and how the presence of it makes the soul of any creature to be absolutely fearless and indomitable?

I have.

Then now we have a clear notion of the bodily qualities which are required in the guardian.

True.

And also of the mental ones; his soul is to be full of spirit?

Yes.

But are not these spirited natures apt to be savage with one another, and with everybody else?

A difficulty by no means easy to overcome, he replied.

Whereas, I said, they ought to be dangerous to their enemies, and gentle to their friends; if not, they will destroy themselves without waiting for their enemies to destroy them.

True, he said.

What is to be done then? I said; how shall we find a gentle nature which has also a great spirit, for the one is the contradiction of the other?

True.

He will not be a good guardian who is wanting in either of these two qualities; and yet the combination of them appears to be impossible; and hence we must infer that to be a good guardian is impossible.

I am afraid that what you say is true, he replied.

Here feeling perplexed I began to think over what had preceded. – My friend, I said, no wonder that we are in a perplexity; for we have lost sight of the image which we had before us.

What do you mean? he said.

I mean to say that there do exist natures gifted with those opposite qualities.

And where do you find them?

Many animals, I replied, furnish examples of them; our friend the dog is a very good one: you know that well-bred dogs are perfectly gentle to their familiars and acquaintances, and the reverse to strangers.

Yes, I know.

Then there is nothing impossible or out of the order of nature in our finding a guardian who has a similar combination of qualities?

Certainly not.

Would not he who is fitted to be a guardian, besides the spirited nature, need to have the qualities of a philosopher?

[376] I do not apprehend your meaning.

The trait of which I am speaking, I replied, may be also seen in the dog, and is remarkable in the animal.

What trait?

Why, a dog, whenever he sees a stranger, is angry; when an acquaintance, he welcomes him, although the one has never done him any harm, nor the other any good. Did this never strike you as curious?

The matter never struck me before; but I quite recognize the truth of your remark.

And surely this instinct of the dog is very charming; – your dog is a true philosopher.

Why?

Why, because he distinguishes the face of a friend and of an enemy only by the criterion of knowing and not knowing. And must not an animal be a lover of learning who determines what he likes and dislikes by the test of knowledge and ignorance?

Most assuredly.

And is not the love of learning the love of wisdom, which is philosophy?

They are the same, he replied.

And may we not say confidently of man also, that he who is likely to be gentle to his friends and acquaintances, must by nature be a lover of wisdom and knowledge?

That we may safely affirm.

Then he who is to be a really good and noble guardian of the State will require to unite in himself philosophy and spirit and swiftness and strength?

Undoubtedly.

Then we have found the desired natures; and now that we have found them, how are they to be reared and educated? Is not this an enquiry which may be expected to throw light on the greater enquiry which is our final end – How do justice and injustice grow up in States? For we do not want either to omit what is to the point or to draw out the argument to an inconvenient length.

Adeimantus thought that the enquiry would be of great service to us.

Then, I said, my dear friend, the task must not be given up, even if somewhat long.

Certainly not.

Come then, and let us pass a leisure hour in story-telling, and our story shall be the education of our heroes.

By all means.

And what shall be their education? Can we find a better than the traditional sort? – And this has two divisions: gymnastic for the body, and music for the soul.

True.

Shall we begin education with music, and go on to gymnastic afterwards?

By all means.

And when you speak of music, do you include literature or not?

I do.

And literature may be either true or false?

Yes.

[377] And the young should be trained in both kinds, and we begin with the false?

I do not understand your meaning, he said.

You know, I said, that we begin by telling children stories which, though not wholly destitute of truth, are in the main fictitious; and these stories are told them when they are not of an age to learn gymnastics.

Very true.

That was my meaning when I said that we must teach music before gymnastics.

Quite right, he said.

You know also that the beginning is the most important part of any work, especially in the case of a young and tender thing; for that is the time at which the character is being formed and the desired impression is more readily taken.

Quite true.

And shall we just carelessly allow children to hear any casual tales which may be devised by casual persons, and to receive into their minds ideas, for the most part, the very opposite of those which we should wish them to have when they are grown up?

We cannot.

Then the first thing will be to establish a censorship of the writers of fiction, and let the censors receive any tale of fiction which is good, and reject the bad; and we will desire mothers and nurses to tell their children the authorised ones only. Let them fashion the mind with such tales, even more fondly than they mould the body with their hands; but most of those which are now in use must be discarded.

Of what tales are you speaking? he said.

You may find a model of the lesser in the greater, I said; for they are necessarily of the same type, and there is the same spirit in both of them.

Very likely, he replied; but I do not as yet know what you would term the greater.

Those, I said, which are narrated by Homer and Hesiod, and the rest of the poets, who have ever been the great story-tellers of mankind.

But which stories do you mean, he said; and what fault do you find with them?

A fault which is most serious, I said; the fault of telling a lie, and, what is more, a bad lie.

But when is this fault committed?

Whenever an erroneous representation is made of the nature of gods and heroes, – as when a painter paints a portrait not having the shadow of a likeness to the original.

Yes, he said, that sort of thing is certainly very blameable; but what are the stories which you mean?

First of all, I said, there was that greatest of all lies in high places, which the poet told about Uranus, and which was a bad lie too, – I mean what Hesiod says that Uranus did, and how Cronus [378] retaliated on him. The doings of Cronus, and the sufferings which in turn his son inflicted upon him, even if they were true, ought certainly not to be lightly told to young and thoughtless persons; if possible, they had better be buried in silence. But if there is an absolute necessity for their mention, a chosen few might hear them in a mystery, and they should sacrifice not a common (Eleusinian) pig, but some huge and unprocurable victim; and then the number of the hearers will be very few indeed.

Why, yes, said he, those stories are extremely objectionable.

Yes, Adeimantus, they are stories not to be repeated in our State; the young man should not be told that in committing the worst of crimes he is far from doing anything outrageous; and that even if he chastises his father when he does wrong, in whatever manner, he will only be following the example of the first and greatest among the gods.

I entirely agree with you, he said; in my opinion those stories are quite unfit to be repeated.

Neither, if we mean our future guardians to regard the habit of quarrelling among themselves as of all things the basest, should any word be said to them of the wars in heaven, and of the plots and fightings of the gods against one another, for they are not true. No, we shall never mention the battles of the giants, or let them be embroidered on garments; and we shall be silent about the innumerable other quarrels of gods and heroes with their friends and relatives. If they would only believe us we would tell them that quarrelling is unholy, and that never up to this time has there been any quarrel between citizens; this is what old men and old women should begin by telling children;

and when they grow up, the poets also should be told to compose for them in a similar spirit. But the narrative of Hephaestus binding here his mother, or how on another occasion Zeus sent him flying for taking her part when she was being beaten, and all the battles of the gods in Homer – these tales must not be admitted into our State, whether they are supposed to have an allegorical meaning or not. For a young person cannot judge what is allegorical and what is literal; anything that he receives into his mind at that age is likely to become indelible and unalterable; and therefore it is most important that the tales which the young first hear should be models of virtuous thoughts.

There you are right, he replied; but if anyone asks where are such models to be found and of what tales are you speaking – how shall we answer him?

I said to him, you and I, Adeimantus, [379] at this moment are not poets, but founders of a State: now the founders of a State ought to know the general forms in which poets should cast their tales, and the limits which must be observed by them, but to make the tales is not their business.

Very true, he said; but what are these forms of theology which you mean?

Something of this kind, I replied: – God is always to be represented as he truly is, whatever be the sort of poetry, epic, lyric or tragic, in which the representation is given.

Right.

And is he not truly good? And must he not be represented as such?

Certainly.

And no good thing is hurtful?

No, indeed.

And that which is not hurtful hurts not?

Certainly not.

And that which hurts not does no evil?

No.

And can that which does no evil be a cause of evil?

Impossible.

And the good is advantageous?

Yes.

And therefore the cause of well-being?

Yes.

It follows therefore that the good is not the cause of all things, but of the good only?

Assuredly.

Then God, if he be good, is not the author of all things, as the many assert, but he is the cause of a few things only, and not of most things that occur to men. For few are the goods of human life, and many are the evils, and the good is to be attributed to God alone; of the evils the causes are to be sought elsewhere, and not in him.

That appears to me to be most true, he said.

Then we must not listen to Homer or to any other poet who is guilty of the folly of saying that two casks

'Lie at the threshold of Zeus, full of lots, one of good, the other of evil lots,'

And that he to whom Zeus gives a mixture of the two,

'Sometimes meets with evil fortune, at other times with good,'

But that he to whom is given the cup of unmingled ill,

'Him wild hunger drives o'er the beauteous earth.'

And again –

'Zeus, who is the dispenser of good and evil to us.'

And if anyone asserts that the violation of oaths and treaties, which was really the work of Pandarus, was brought about by Athene and Zeus, [380] or that the strife and contention of the gods was instigated by Themis and Zeus, he shall not have our approval; neither will we allow our young men to hear the words of Aeschylus, that,

'God plants guilt among men when he desires utterly to destroy a house.'

And if a poet writes of the sufferings of Niobe – the subject of the tragedy in which these iambic verses occur – or of the house of Pelops, or of the Trojan war or on any similar theme, either we must not permit him to say that these are the works of God, or if they are of God, he must devise some explanation of them such as we are seeking; he must say that God did what was just and right, and they were the better for being punished; but that those who are punished are miserable, and that God is the author of their misery – the poet is not to be permitted to say; though he may say that the wicked are miserable because they require to be punished, and are benefited by receiving punishment from God; but that God being good is the author of evil to anyone is to be strenuously denied, and not to be said or sung or heard in verse or prose by anyone whether old or young in any well-ordered commonwealth. Such a fiction is suicidal, ruinous, impious.

I agree with you, he replied, and am ready to give my assent to the law.

Let this then be one of our rules and principles concerning the gods, to which our poets and reciters will be expected to conform, – that God is not the author of all things, but of good only.

That will do, he said.

And what do you think of a second principle? Shall I ask you whether God is a magician, and of a nature to appear insidiously now in one shape, and now in another – sometimes himself changing and passing into many forms, sometimes deceiving us with the semblance of such transformations; or is he one and the same immutably fixed in his own proper image?

I cannot answer you, he said, without more thought.

Well, I said; but if we suppose a change in anything, that change must be effected either by the thing itself, or by some other thing?

Most certainly.

And things which are at their best are also least liable to be altered or discomposed; for example, when healthiest and strongest, the human frame is least liable to be affected by meats and drinks, and the plant which is in the fullest vigour also suffers least from winds or the heat of the sun or any similar causes.

[381] Of course.

And will not the bravest and wisest soul be least confused or deranged by any external influence?

True.

And the same principle, as I should suppose, applies to all composite things – furniture, houses, garments: when good and well made, they are least altered by time and circumstances.

Very true.

Then everything which is good, whether made by art or nature, or both, is least liable to suffer change from without?

True.

But surely God and the things of God are in every way perfect?

Of course they are.

Then he can hardly be compelled by external influence to take many shapes?

He cannot.

But may he not change and transform himself?

Clearly, he said, that must be the case if he is changed at all.

And will he then change himself for the better and fairer, or for the worse and more unsightly?

If he change at all he can only change for the worse, for we cannot suppose him to be deficient either in virtue or beauty.

Very true, Adeimantus; but then, would anyone, whether God or man, desire to make himself worse?

Impossible.

Then it is impossible that God should ever be willing to change; being, as is supposed, the fairest and best that is conceivable, every God remains absolutely and forever in his own form.

That necessarily follows, he said, in my judgment.

[minor omission, 381d-e]

But although the gods are themselves unchangeable, still by witchcraft and deception they may make us think that they appear in various forms?

Perhaps, he replied.

[382] Well, but can you imagine that God will be willing to lie, whether in word or deed, or to put forth a phantom of himself?

I cannot say, he replied.

Do you not know, I said, that the true lie, if such an expression may be allowed, is hated of gods and men?

What do you mean? he said.

I mean that no one is willingly deceived in that which is the truest and highest part of himself, or about the truest and highest matters; there, above all, he is most afraid of a lie having possession of him.

Still, he said, I do not comprehend you.

The reason is, I replied, that you attribute some profound meaning to my words; but I am only saying that deception, or being deceived or uninformed about the highest realities in the highest part of themselves, which is the soul, and in that part of them to have and to hold the lie, is what mankind least likes; – that, I say, is what they utterly detest.

There is nothing more hateful to them.

And, as I was just now remarking, this ignorance in the soul of him who is deceived may be called the true lie; for the lie in words is only a kind of imitation and shadowy image of a previous affection of the soul, not pure unadulterated falsehood. Am I not right?

Perfectly right.

[minor omission, 382a-e]

Then is God perfectly simple and true both in word and deed; he changes not; he deceives not, either by sign or word, by dream or waking vision.

Your thoughts, he said, are the reflection of my own.

You agree with me then, I said, that this is the second type or form in which we should write and speak about divine things. The gods are not magicians who transform themselves, neither do they deceive mankind in any way.

I grant that.

[minor omission, 383a-b]

These are the kind of sentiments about the gods which will arouse our anger; and he who utters them shall be refused a chorus; neither shall we allow teachers to make use of them in the instruction of the young, meaning, as we do, that our guardians, as far as men can be, should be true worshippers of the gods and like them.

I entirely agree, he said, in these principles, and promise to make them my laws.

BOOK III

[386a]

Such then, I said, are our principles of theology – some tales are to be told, and others are not to be told to our disciples from their youth upwards, if we mean them to honour the gods and their parents, and to value friendship with one another.

Yes; and I think that our principles are right, he said.

But if they are to be courageous, must they not learn other lessons besides these, and lessons of such a kind as will take away the fear of death? Can any man be courageous who has the fear of death in him?

Certainly not, he said.

And can he be fearless of death, or will he choose death in battle rather than defeat and slavery, who believes the world below to be real and terrible?

Impossible.

Then we must assume a control over the narrators of this class of tales as well as over the others, and beg them not simply to revile but rather to commend the world below, intimating to them that their descriptions are untrue, and will do harm to our future warriors.

That will be our duty, he said.

Then, I said, we shall have to obliterate many obnoxious passages, beginning with the verses,

'I would rather be a serf on the land of a poor and portionless man than rule over all the dead who have come to nought.'

[examples omitted, 386c-387a]

'As bats in hollow of mystic cavern, whenever any of them has dropped out of the string and falls from the rock, fly shrilling and cling to one another, so did they with shrilling cry hold together as they moved.'

And we must beg Homer and the other poets not to be angry if we strike out these and similar passages, not because they are unpoetical, or unattractive to the popular ear, but because the greater the poetical charm of them, the less are they meet for the ears of boys and men who are meant to be free, and who should fear slavery more than death.

Undoubtedly.

Also we shall have to reject all the terrible and appalling names which describe the world below – Cocytus and Styx, ghosts under the earth, and sapless shades,

[minor omission, 387c]

And shall we proceed to get rid of the weepings and wailings of famous men?

They will go with the rest.

But shall we be right in getting rid of them? Reflect: our principle is that the good man will not consider death terrible to any other good man who is his comrade.

Yes; that is our principle.

And therefore he will not sorrow for his departed friend as though he had suffered anything terrible?

He will not.

Such a one, as we further maintain, is sufficient for himself and his own happiness, and therefore is least in need of other men.

True, he said.

And for this reason the loss of a son or brother, or the deprivation of fortune, is to him of all men least terrible.

Assuredly.

And therefore he will be least likely to lament, and will bear with the greatest equanimity any misfortune of this sort which may befall him.

Yes, he will feel such a misfortune far less than another.

Then we shall be right in getting rid of the lamentations of famous men, and making them over to women (and not even to women who are good for anything), or to men of a baser sort, that those who are being educated by us to be the defenders of their country may scorn to do the like.

That will be very right.

Then we will once more entreat Homer and the other poets not to depict Achilles, who is the son of a goddess, first lying on his side, then on his back, and then on his face; then starting up and sailing in a frenzy along the shores of the barren sea;

[minor omissions, 388b-e]

But if he must introduce the gods, at any rate let him not dare so completely to misrepresent the greatest of the gods, as to make him say –

'O heavens! With my eyes verily I behold a dear friend of mine chased round and round the city, and my heart is sorrowful.'

[...]

For if, my sweet Adeimantus, our youth seriously listen to such unworthy representations of the gods, instead of laughing at them as they ought, hardly will any of them deem that he himself, being but a man, can be dishonoured by similar actions; neither will he rebuke any inclination which may arise in his mind to say and do the like. And instead of having any shame or self-control, he will be always whining and lamenting on slight occasions.

Yes, he said, that is most true.

[...]

Neither ought our guardians to be given to laughter. For a fit of laughter which has been indulged to excess almost always produces a violent reaction.

So I believe.

Then persons of worth, even if only mortal men, must not be represented as overcome [389] by laughter, and still less must such a representation of the gods be allowed.

Still less of the gods, as you say, he replied.

Then we shall not suffer such an expression to be used about the gods as that of Homer when he describes how,

'Inextinguishable laughter arose among the blessed gods, when they saw Hephaestus bustling about the mansion.'

On your views, we must not admit them.

On my views, if you like to father them on me; that we must not admit them is certain.

Again, truth should be highly valued; if, as we were saying, a lie is useless to the gods, and useful only as a medicine to men, then the use of such medicines should be restricted to physicians; private individuals have no business with them.

Clearly not, he said.

Then if anyone at all is to have the privilege of lying, the rulers of the State should be the persons; and they, in their dealings either with enemies or with their own citizens, may be allowed to lie for the public good. But nobody else should meddle with anything of the kind; and although the rulers have this privilege, for a private man to lie to them in return is to be deemed a more heinous fault than for the patient or the pupil of a gymnasium not to speak the truth about his own bodily illnesses to the physician or to the trainer, or for a sailor not to tell the captain what is happening about the ship and the rest of the crew, and how things are going with himself or his fellow sailors.

Most true, he said.

If, then, the ruler catches anybody beside himself lying in the State,

'Any of the craftsmen, whether he be priest or physician or carpenter,'

He will punish him for introducing a practice which is equally subversive and destructive of ship or State.

Most certainly, he said, if our idea of the State is ever carried out.

In the next place our youth must be temperate?

Certainly.

Are not the chief elements of temperance, speaking generally, obedience to commanders and self-control in sensual pleasures?

True.

Then we shall approve such language as that of Diomede in Homer,

'Friend, sit still and obey my word,'

[390] And the verses which follow,

'The Greeks marched breathing prowess, ...in silent awe of their leaders,'

And other sentiments of the same kind.

We shall.

What of this line,

'O heavy with wine, who hast the eyes of a dog and the heart of a stag,'

And of the words which follow? Would you say that these, or any similar impertinences which private individuals are supposed to address to their rulers, whether in verse or prose, are well or ill spoken?

They are ill spoken.

They may very possibly afford some amusement, but they do not conduce to temperance. And therefore they are likely to do harm to our young men – you would agree with me there?

Yes.

And then, again, to make the wisest of men say that nothing in his opinion is more glorious than,

'When the tables are full of bread and meat, and the cup-bearer carries round wine which he draws from the bowl and pours into the cups,'

Is it fit or conducive to temperance for a young man to hear such words? Or the verse,

'The saddest of fates is to die and meet destiny from hunger?'

What would you say again to the tale of Zeus, who, while other gods and men were asleep and he the only person awake, lay devising plans, but forgot them all in a moment through his lust, and was so completely overcome at the sight of Hera that he would not even go into the hut, but wanted to lie with her on the ground, declaring that he had never been in such a state of rapture before, even when they first met one another,

'Without the knowledge of their parents,'

Or that other tale of how Hephaestus, because of similar goings on, cast a chain around Ares and Aphrodite?

Indeed, he said, I am strongly of opinion that they ought not to hear that sort of thing.

But any deeds of endurance which are done or told by famous men, these they ought to see and hear; as, for example, what is said in the verses,

'He smote his breast, and thus reproached his heart, Endure, my heart; far worse hast thou endured!'

Certainly, he said.

In the next place, we must not let them be receivers of gifts or lovers of money.

Certainly not.

Neither must we sing to them of,

'Gifts persuading gods, and persuading reverend kings.'

Neither is Phoenix, the tutor of Achilles, to be approved or deemed to have given his pupil good counsel when he told him that he should take the gifts of the Greeks and assist them; but that without a gift he should not lay aside his anger. Neither will we believe or acknowledge Achilles himself to have been such a lover of money that he [391] took Agamemnon's gifts, or that when he had received payment he restored the dead body of Hector, but that without payment he was unwilling to do so.

Undoubtedly, he said, these are not sentiments which can be approved.

[minor omission, 391a-c]

And let us equally refuse to believe, or allow to be repeated, the tale of Theseus son of Poseidon, or of Peirithous son of Zeus, going forth as they did to perpetrate a horrid rape; or of any other hero or son of a god daring to do such impious and dreadful things as they falsely ascribe to them in our day.

[minor omission, 391d-392a]

The manner in which gods and demigods and heroes and the world below should be treated has been already laid down.

Very true.

And what shall we say about men? That is clearly the remaining portion of our subject.

Clearly so.

[minor omission, 391a-b]

That such things are or are not to be said about men is a question which we cannot determine until we have discovered what justice is, and how naturally advantageous to the possessor, whether he seem to be just or not.

Most true, he said.

Enough of the subjects of poetry: let us now speak of the style; and when this has been considered, both matter and manner will have been completely treated.

I do not understand what you mean, said Adeimantus.

Then I must make you understand; and perhaps I may be more intelligible if I put the matter in this way. You are aware, I suppose, that all mythology and poetry is a narration of events, either past, present, or to come?

Certainly, he replied.

And narration may be either simple narration, or imitation, or a union of the two?

That again, he said, I do not quite understand.

I fear that I must be a ridiculous teacher when I have so much difficulty in making myself apprehended. Like a bad speaker, therefore, I will not take the whole of the subject, but will break a piece off in illustration of my meaning. You know the first lines of the Iliad, in which the poet says that Chryses prayed Agamemnon to release his daughter, and that Agamemnon flew into a passion with him; whereupon Chryses, failing of his object, [393] invoked the anger of the God against the Achaeans. Now as far as these lines,

'And he prayed all the Greeks, but especially the two sons of Atreus, the chiefs of the people,'

The poet is speaking in his own person; he never leads us to suppose that he is anyone else. But in what follows he takes the person of Chryses, and then he does all that he can to make us believe that the speaker is not Homer, but the aged priest himself. And in this double form he has cast the entire narrative of the events which occurred at Troy and in Ithaca and throughout the Odyssey.

Yes.

And a narrative it remains both in the speeches which the poet recites from time to time and in the intermediate passages?

Quite true.

But when the poet speaks in the person of another, may we not say that he assimilates his style to that of the person who, as he informs you, is going to speak?

Certainly.

And this assimilation of himself to another, either by the use of voice or gesture, is the imitation of the person whose character he assumes?

Of course.

Then in this case the narrative of the poet may be said to proceed by way of imitation?

Very true.

Or, if the poet everywhere appears and never conceals himself, then again the imitation is dropped, and his poetry becomes simple narration.

[minor omission, 393d-394b]

I understand, he said.

Or you may suppose the opposite case – that the intermediate passages are omitted, and the dialogue only left.

That also, he said, I understand; you mean, for example, as in tragedy.

You have conceived my meaning perfectly; and if I mistake not, what you failed to apprehend before is now made clear to you, that poetry and mythology are, in some cases, wholly imitative – instances of this are supplied by tragedy and comedy; there is likewise the opposite style, in which the poet is the only speaker – of this the dithyramb affords the best example; and the combination of both is found in epic, and in several other styles of poetry. Do I take you with me?

Yes, he said; I see now what you meant.

I will ask you to remember also what I began by saying, that we had done with the subject and might proceed to the style.

Yes, I remember.

In saying this, I intended to imply that we must come to an understanding about the mimetic art, – whether the poets, in narrating their stories, are to be allowed by us to imitate, and if so, whether in whole or in part, and if the latter, in what parts; or should all imitation be prohibited?

You mean, I suspect, to ask whether tragedy and comedy shall be admitted into our State?

Yes, I said; but there may be more than this in question: I really do not know as yet, but whither the argument may blow, thither we go.

And go we will, he said.

Then, Adeimantus, let me ask you whether our guardians ought to be imitators; or rather, has not this question been decided by the rule already laid down, that one man can only do one thing well, and not many; and that if he attempt many, he will altogether fail of gaining much reputation in any?

Certainly.

And this is equally true of imitation; no one man can imitate many things as well as he would imitate a single one?

He cannot.

[395] Then the same person will hardly be able to play a serious part in life, and at the same time to be an imitator and imitate many other parts as well; for even when two species of imitation are nearly allied, the same persons cannot succeed in both, as, for example, the writers of tragedy and comedy – did you not just now call them imitations?

Yes, I did; and you are right in thinking that the same persons cannot succeed in both.

Any more than they can be rhapsodists and actors at once?

True.

Neither are comic and tragic actors the same; yet all these things are but imitations.

They are so.

And human nature, Adeimantus, appears to have been coined into yet smaller pieces, and to be as incapable of imitating many things well, as of performing well the actions of which the imitations are copies.

Quite true, he replied.

If then we adhere to our original notion and bear in mind that our guardians, setting aside every other business, are to dedicate themselves wholly to the maintenance of freedom in the State, making this their craft, and engaging in no work which does not bear on this end, they ought not to practise or imitate anything else; if they imitate at all, they should imitate from youth upward only those characters which are suitable to their profession – the courageous, temperate, holy, free, and the like; but they should not depict or be skilful at imitating any kind of illiberality or baseness, lest from imitation they should come to be what they imitate. Did you never observe how imitations, beginning in early youth and continuing far into life, at length grow into habits and become a second nature, affecting body, voice, and mind?

Yes, certainly, he said.

Then, I said, we will not allow those for whom we profess a care and of whom we say that they ought to be good men, to imitate a woman, whether young or old, quarrelling with her husband, or striving and vaunting against the gods in conceit of her happiness, or when she is in affliction, or sorrow, or weeping; and certainly not one who is in sickness, love, or labour.

Very right, he said.

Neither must they represent slaves, male or female, performing the offices of slaves?

They must not.

And surely not bad men, whether cowards or any others, who do the reverse of what we have just been prescribing, who scold or mock or revile one another in drink [395] or out of drink, or who in any other manner sin against themselves and their neighbours in word or deed, as the manner of such is. Neither should they be trained to imitate the action or speech of men or women who are mad or bad; for madness, like vice, is to be known but not to be practised or imitated.

[minor omission, 396a-b]

You mean, I said, if I understand you aright, that there is one sort of narrative style which may be employed by a truly good man when he has anything to say, and that another sort will be used by a man of an opposite character and education.

And which are these two sorts? he asked.

Suppose, I answered, that a just and good man in the course of a narration comes on some saying or action of another good man, – I should imagine that he will like to personate him, and will not be ashamed of this sort of imitation: he will be most ready to play the part of the good man when he is acting firmly and wisely; in a less degree when he is overtaken by illness or love or drink, or has met with any other disaster. But when he comes to a character which is unworthy of him, he will not make a study of that; he will disdain such a person, and will assume his likeness, if at all, for a moment only when he is performing some good action; at other times he will be ashamed to play a part which he has never practised, nor will he like to fashion and frame himself after the baser models; he feels the employment of such an art, unless in jest, to be beneath him, and his mind revolts at it.

So I should expect, he replied.

Then he will adopt a mode of narration such as we have illustrated out of Homer, that is to say, his style will be both imitative and narrative; but there will be very little of the former, and a great deal of the latter. Do you agree?

Certainly, he said; that is the model which such a speaker must necessarily take.

[397] But there is another sort of character who will narrate anything, and, the worse he is, the more unscrupulous he will be; nothing will be too bad for him: and he will be ready to imitate anything, not as a joke, but in right good earnest, and before a large company. As I was just now saying, he will attempt to represent the roll of thunder, the noise of wind and hail, or the creaking of wheels, and pulleys, and the various sounds of flutes, pipes, trumpets, and all sorts of instruments: he will bark like a dog, bleat like a sheep, or crow like a cock; his entire art will consist in imitation of voice and gesture, and there will be very little narration.

That, he said, will be his mode of speaking.

These, then, are the two kinds of style?

Yes.

And you would agree with me in saying that one of them is simple and has but slight changes; and if the harmony and rhythm are also chosen for their simplicity, the result is that the speaker, if he speaks correctly, is always pretty much the same in style, and he will keep within the limits of a single harmony (for the changes are not great), and in like manner he will make use of nearly the same rhythm?

That is quite true, he said.

Whereas the other requires all sorts of harmonies and all sorts of rhythms, if the music and the style are to correspond, because the style has all sorts of changes.

That is also perfectly true, he replied.

And do not the two styles, or the mixture of the two, comprehend all poetry, and every form of expression in words? No one can say anything except in one or other of them or in both together.

They include all, he said.

And shall we receive into our State all the three styles, or one only of the two unmixed styles? Or would you include the mixed?

I should prefer only to admit the pure imitator of virtue.

[minor omission, 397d-398b]

Then now, my friend, I said, that part of music or literary education which relates to the story or myth may be considered to be finished; for the matter and manner have both been discussed.

I think so too, he said.

Next in order will follow melody and song.

That is obvious.

Everyone can see already what we ought to say about them, if we are to be consistent with ourselves.

I fear, said Glaucon, laughing, that the word 'everyone' hardly includes me, for I cannot at the moment say what they should be; though I may guess.

At any rate you can tell that a song or ode has three parts – the words, the melody, and the rhythm; that degree of knowledge I may presuppose?

Yes, he said; so much as that you may.

And as for the words, there will surely be no difference between words which are and which are not set to music; both will conform to the same laws, and these have been already determined by us?

Yes.

And the melody and rhythm will depend upon the words?

Certainly.

We were saying, when we spoke of the subject-matter, that we had no need of lamentation and strains of sorrow?

True.

And which are the harmonies expressive of sorrow? You are musical, and can tell me.

The harmonies which you mean are the mixed or tenor Lydian, and the full-toned or bass Lydian, and such like.

These then, I said, must be banished; even to women who have a character to maintain they are of no use, and much less to men.

Certainly.

In the next place, drunkenness and softness and indolence are utterly unbecoming the character of our guardians.

Utterly unbecoming.

And which are the soft or drinking harmonies?

The Ionian, he replied, and the Lydian; they are termed 'relaxed.'

[399] Well, and are these of any military use?

Quite the reverse, he replied; and if so the Dorian and the Phrygian are the only ones which you have left.

[minor omission, 399a-e]

Then, I said, if these and these only are to be used in our songs and melodies, we shall not want multiplicity of notes or a panharmonic scale?

I suppose not.

Then we shall not maintain the artificers of lyres with three corners and complex scales, or the makers of any other many-stringed curiously-harmonised instruments?

Certainly not.

But what do you say to flute-makers and flute-players? Would you admit them into our State when you reflect that in this composite use of harmony the flute is worse than all the stringed instruments put together; even the panharmonic music is only an imitation of the flute?

Clearly not.

There remain then only the lyre and the harp for use in the city, and the shepherds may have a pipe in the country.

That is surely the conclusion to be drawn from the argument.

The preferring of Apollo and his instruments to Marsyas and his instruments is not at all strange, I said.

Not at all, he replied.

And so, by the dog, we have been unconsciously purging the State, which not long ago we termed luxurious.

And we have done wisely, he replied.

Then let us now finish the purgation, I said. Next in order to harmonies, rhythms will naturally follow, and they should be subject to the same rules, for we ought not to seek out complex systems of metre, or metres of every kind, but rather to discover what rhythms are the expressions of a courageous and harmonious life.

[minor omissions, 399-400c]

We must take Damon into our counsels; and he will tell us what rhythms are expressive of meanness, or insolence, or fury, or other unworthiness, and what are to be reserved for the expression of opposite feelings.

[...]

Rather so, I should say.

But there is no difficulty in seeing that grace or the absence of grace is an effect of good or bad rhythm.

None at all.

And also that good and bad rhythm naturally assimilate to a good and bad style; and that harmony and discord in like manner follow style; for our principle is that rhythm and harmony are regulated by the words, and not the words by them.

Just so, he said, they should follow the words.

And will not the words and the character of the style depend on the temper of the soul?

Yes.

And everything else on the style?

Yes.

Then beauty of style and harmony and grace and good rhythm depend on simplicity, – I mean the true simplicity of a rightly and nobly ordered mind and character, not that other simplicity which is only an euphemism for folly?

Very true, he replied.

And if our youth are to do their work in life, must they not make these graces and harmonies their perpetual aim?

They must.

[401] And surely the art of the painter and every other creative and constructive art are full of them, – weaving, embroidery, architecture, and every kind of manufacture; also nature, animal and vegetable, – in all of them there is grace or the absence of grace. And ugliness and discord and inharmonious motion are nearly allied to ill words and ill nature, as grace and harmony are the twin sisters of goodness and virtue and bear their likeness.

That is quite true, he said.

But shall our superintendence go no further, and are the poets only to be required by us to express the image of the good in their works, on pain, if they do anything else, of expulsion from our State? Or is the same control to be extended to other artists, and are they also to be prohibited from exhibiting the opposite forms of vice and intemperance and meanness and indecency in sculpture and building and the other creative arts; and is he who cannot conform to this rule of ours to be prevented from practising his art in our State, lest the taste of our citizens be corrupted by him? We would not have our guardians grow up amid images of moral deformity, as in some noxious pasture, and there browse and feed upon many a baneful herb and flower day by day, little by little, until they silently gather a festering mass of corruption in their own soul. Let our artists rather be those who are gifted to discern the true nature of the beautiful and graceful; then will our youth dwell in a land of health, amid fair sights and sounds, and receive the good in everything; and beauty, the effluence of fair works, shall flow into the eye and ear, like a health-giving breeze from a purer region, and insensibly draw the soul from earliest years into likeness and sympathy with the beauty of reason.

There can be no nobler training than that, he replied.

And therefore, I said, Glaucon, musical training is a more potent instrument than any other, because rhythm and harmony find their way into the inward places of the soul, on which they mightily fasten, imparting grace, and making the soul of him who is rightly educated graceful, or of him who is ill-educated ungraceful; and also because he who has received this true education of the inner being will most shrewdly perceive omissions or faults in art and nature, and with a true taste, while he praises and rejoices over and receives into his soul the good, [402] and becomes noble and good, he will justly blame and hate the bad, now in the days of his youth, even before he is able to know the reason why; and when reason comes he will recognize and salute the friend with whom his education has made him long familiar.

Yes, he said.

[minor omission, 402a-b]

Even so, as I maintain, neither we nor our guardians, whom we have to educate, can ever become musical until we and they know the essential forms of temperance, courage, liberality, magnificence, and their kindred, as well as the contrary forms, in all their combinations, and can recognize them and their images wherever they are found, not slighting them either in small things or great, but believing them all to be within the sphere of one art and study.

Most assuredly.

And when a beautiful soul harmonizes with a beautiful form, and the two are cast in one mould, that will be the fairest of sights to him who has an eye to see it?

The fairest indeed.

And the fairest is also the loveliest?

That may be assumed.

And the man who has the spirit of harmony will be most in love with the loveliest; but he will not love him who is of an inharmonious soul?

That is true, he replied, if the deficiency be in his soul; but if there be any merely bodily defect in another he will be patient of it, and will love all the same.

I perceive, I said, that you have or have had experiences of this sort, and I agree. But let me ask you another question: Has excess of pleasure any affinity to temperance?

How can that be? he replied; pleasure deprives a man of the use of his faculties quite as much as pain.

Or any affinity to virtue in general?

[403] None whatever.

Any affinity to wantonness and intemperance?

Yes, the greatest.

And is there any greater or keener pleasure than that of sensual love?

No, nor a madder.

Whereas true love is a love of beauty and order – temperate and harmonious?

Quite true, he said.

Then no intemperance or madness should be allowed to approach true love?

Certainly not.

Then mad or intemperate pleasure must never be allowed to come near the lover and his beloved.

[minor omission, 403b-c]

Thus much of music, which makes a fair ending; for what should be the end of music if not the love of beauty?

I agree, he said.

After music comes gymnastic, in which our youth are next to be trained.

Certainly.

Gymnastic as well as music should begin in early years; the training in it should be careful and should continue through life. Now my belief is ... not that the good body by any bodily excellence improves the soul, but, on the contrary, that the good soul, by her own excellence, improves the body as far as this may be possible. What do you say?

Yes, I agree.

Then, to the mind when adequately trained, we shall be right in handing over the more particular care of the body; and in order to avoid prolixity we will now only give the general outlines of the subject.

Very good.

That they must abstain from intoxication has been already remarked by us; for of all persons a guardian should be the last to get drunk and not know where in the world he is.

Yes, he said; that a guardian should require another guardian to take care of him is ridiculous indeed.

But next, what shall we say of their food; for the men are in training for the great contest of all – are they not?

Yes, he said.

[404] And will the habit of body of our ordinary athletes be suited to them?

Why not?

I am afraid, I said, that a habit of body such as they have is but a sleepy sort of thing, and rather perilous to health. Do you not observe that these athletes sleep away their lives, and are liable to most dangerous illnesses if they depart, in ever so slight a degree, from their customary regimen?

Yes, I do.

Then, I said, a finer sort of training will be required for our warrior athletes, who are to be like wakeful dogs, and to see and hear with the utmost keenness; amid the many changes of water and also of food, of summer heat and winter cold, which they will have to endure when on a campaign, they must not be liable to break down in health.

That is my view.

The really excellent gymnastic is twin sister of that simple music which we were just now describing.

How so?

Why, I conceive that there is a gymnastic which, like our music, is simple and good; and especially the military gymnastic.

What do you mean?

My meaning may be learned from Homer; he, you know, feeds his heroes at their feasts, when they are campaigning, on soldiers' fare; they have no fish, although they are on the shores of the Hellespont, and they are not allowed boiled meats but only roast, which is the food most convenient for soldiers, requiring only that they should light a fire, and not involving the trouble of carrying about pots and pans.

True.

And I can hardly be mistaken in saying that sweet sauces are nowhere mentioned in Homer. In proscribing them, however, he is not singular; all professional athletes are well aware that a man who is to be in good condition should take nothing of the kind.

Yes, he said; and knowing this, they are quite right in not taking them.

Then you would not approve of Syracusan dinners, and the refinements of Sicilian cookery?

I think not.

Nor, if a man is to be in condition, would you allow him to have a Corinthian girl as his fair friend?

Certainly not.

Neither would you approve of the delicacies, as they are thought, of Athenian confectionary?

Certainly not.

All such feeding and living may be rightly compared by us to melody and song composed in the panharmonic style, and in all the rhythms.

Exactly.

There complexity engendered licence, and here disease; whereas simplicity in music was the parent of temperance in the soul; and simplicity in gymnastic of health in the body.

Most true, he said.

[405] But when intemperance and diseases multiply in a State, halls of justice and medicine are always being opened; and the arts of the doctor and the lawyer give themselves airs, finding how keen is the interest which not only the slaves but the freemen of a city take about them.

Of course.

And yet what greater proof can there be of a bad and disgraceful state of education than this, that not only artisans and the meaner sort of people need the skill of first-rate physicians and judges, but also those who would profess to have had a liberal education?

[405b-406c omitted]

When a carpenter is ill, he asks the physician for a rough and ready cure; an emetic or a purge or a cautery or the knife, – these are his remedies. And if someone prescribes for him a course of dietetics, and tells him that he must swathe and swaddle his head, and all that sort of thing, he replies at once that he has no time to be ill, and that he sees no good in a life which is spent in nursing his disease to the neglect of his customary employment; and therefore bidding good-bye to this sort of physician, he resumes his ordinary habits, and either gets well and lives and does his business, or, if his constitution fails, he dies and has no more trouble.

Yes, he said, and a man in his condition of life ought to use the art of medicine thus far only.

[407] Has he not, I said, an occupation; and what profit would there be in his life if he were deprived of his occupation?

Quite true, he said.

But with the rich man this is otherwise; of him we do not say that he has any specially appointed work which he must perform, if he would live.

He is generally supposed to have nothing to do.

Then you never heard of the saying of Phocylides, that as soon as a man has a livelihood he should practise virtue?

Nay, he said, I think that he had better begin somewhat sooner.

Let us not have a dispute with him about this, I said; but rather ask ourselves: Is the practice of virtue obligatory on the rich man, or can he live without it? And if obligatory on him, then let us raise a further question, whether this dieting of disorders, which is an impediment to the application of the mind in carpentering and the mechanical arts, does not equally stand in the way of the sentiment of Phocylides?

Of that, he replied, there can be no doubt; such excessive care of the body, when carried beyond the rules of gymnastic, is most inimical to the practice of virtue.

Yes, indeed, I replied, and equally incompatible with the management of a house, an army, or an office of state; and, what is most important of all, irreconcileable with any kind of study or thought or self-reflection – there is a constant suspicion that headache and giddiness are to be ascribed to philosophy, and hence all practising or making trial of virtue in the higher sense is absolutely stopped; for a man is always fancying that he is being made ill, and is in constant anxiety about the state of his body.

Yes, likely enough.

And therefore our politic Asclepius may be supposed to have exhibited the power of his art only to persons who, being generally of healthy constitution and habits of life, had a definite ailment; such as these he cured by purges and operations, and bade them live as usual, herein consulting the interests of the State; but bodies which disease had penetrated through and through he would not have attempted to cure by gradual processes of evacuation and infusion: he did not want to lengthen out good-for-nothing lives, or to have weak fathers begetting weaker sons; – if a man was not able to live in the ordinary way he had no business to cure him; for such a cure would have been of no use either to himself, or to the State.

Then, he said, you regard Asclepius as a statesman.

Clearly; and his character is further illustrated by his sons. [408] Note that they were heroes in the days of old and practised the medicines of which I am speaking at the siege of Troy: You will remember how, when Pandarus wounded Menelaus, they,

'Sucked the blood out of the wound, and sprinkled soothing remedies,'

But they never prescribed what the patient was afterwards to eat or drink in the case of Menelaus, any more than in the case of Eurypylus; the remedies, as they conceived, were enough to heal any man who before he was wounded was healthy and regular in his habits; and even though he did happen to drink a posset of Pramnian wine, he might get well all the same. But they would have nothing to do with unhealthy and intemperate subjects, whose lives were of no use either to themselves or others; the art of medicine was not designed for their good, and though they were as rich as Midas, the sons of Asclepius would have declined to attend them.

[minor omission, 408b-c]

All that, Socrates, is excellent; but I should like to put a question to you: Ought there not to be good physicians in a State, and are not the best those who have treated the greatest number of constitutions good and bad? And are not the best judges in like manner those who are acquainted with all sorts of moral natures?

Yes, I said, I too would have good judges and good physicians. But do you know whom I think good?

Will you tell me?

I will, if I can. Let me, however, note that in the same question you join two things which are not the same.

How so? he asked.

Why, I said, you join physicians and judges. Now the most skilful physicians are those who, from their youth upwards, have combined with the knowledge of their art the greatest experience of disease; they had better not be robust in health, and should have had all manner of diseases in their own persons. For the body, as I conceive, is not the instrument with which they cure the body; in that case we could not allow them ever to be or to have been sickly; but they cure the body with the mind, and the mind which has become and is sick can cure nothing.

That is very true, he said.

[409] But with the judge it is otherwise; since he governs mind by mind; he ought not therefore to have been trained among vicious minds, and to have associated with them from youth upwards, and to have gone through the whole calendar of crime, only in order that he may quickly infer the crimes of others as he might their bodily diseases from his own self-consciousness; the honourable mind which is to form a healthy judgment should have had no experience or contamination of evil habits when young. And this is the reason why in youth good men often appear to be simple, and are easily practised upon by the dishonest, because they have no examples of what evil is in their own souls.

Yes, he said, they are far too apt to be deceived.

Therefore, I said, the judge should not be young; he should have learned to know evil, not from his own soul, but from late and long observation of the nature of evil in others: knowledge should be his guide, not personal experience.

Yes, he said, that is the ideal of a judge.

Yes, I replied, and he will be a good man (which is my answer to your question); for he is good who has a good soul. But the cunning and suspicious nature of which we spoke, – he who has committed many crimes, and fancies himself to be a master in wickedness, when he is amongst his fellows, is wonderful in the precautions which he takes, because he judges of them by himself: but when he gets into the company of men of virtue, who have the experience of age, he appears to be a fool again, owing to his unseasonable suspicions; he cannot recognize an honest man, because he has no pattern of honesty in himself; at the same time, as the bad are more numerous than the good, and he meets with them oftener, he thinks himself, and is by others thought to be, rather wise than foolish.

Most true, he said.

Then the good and wise judge whom we are seeking is not this man, but the other; for vice cannot know virtue too, but a virtuous nature, educated by time, will acquire a knowledge both of virtue and vice: the virtuous, and not the vicious, man has wisdom – in my opinion.

And in mine also.

This is the sort of medicine, and this is the sort of law, which you will sanction in your state. They will minister to [410] better natures, giving health both of soul and of body; but those who are diseased in their bodies they will leave to die, and the corrupt and incurable souls they will put an end to themselves.

That is clearly the best thing both for the patients and for the State.

And thus our youth, having been educated only in that simple music which, as we said, inspires temperance, will be reluctant to go to law.

Clearly.

And the musician, who, keeping to the same track, is content to practise the simple gymnastic, will have nothing to do with medicine unless in some extreme case.

That I quite believe.

The very exercises and tolls which he undergoes are intended to stimulate the spirited element of his nature, and not to increase his strength; he will not, like common athletes, use exercise and regimen to develop his muscles.

Very right, he said.

Neither are the two arts of music and gymnastic really designed, as is often supposed, the one for the training of the soul, the other for the training of the body.

What then is the real object of them?

I believe, I said, that the teachers of both have in view chiefly the improvement of the soul.

How can that be? he asked.

Did you never observe, I said, the effect on the mind itself of exclusive devotion to gymnastic, or the opposite effect of an exclusive devotion to music?

In what way shown? he said.

The one producing a temper of hardness and ferocity, the other of softness and effeminacy, I replied.

Yes, he said, I am quite aware that the mere athlete becomes too much of a savage, and that the mere musician is melted and softened beyond what is good for him.

Yet surely, I said, this ferocity only comes from spirit, which, if rightly educated, would give courage, but, if too much intensified, is liable to become hard and brutal.

That I quite think.

On the other hand the philosopher will have the quality of gentleness. And this also, when too much indulged, will turn to softness, but, if educated rightly, will be gentle and moderate.

True.

And in our opinion the guardians ought to have both these qualities?

Assuredly.

And both should be in harmony?

Beyond question.

[411] And the harmonious soul is both temperate and courageous?

Yes.

And the inharmonious is cowardly and boorish?

Very true.

And, when a man allows music to play upon him and to pour into his soul through the funnel of his ears those sweet and soft and melancholy airs of which we were just now speaking, and his whole life is passed in warbling and the delights of song; in the first stage of the process the passion or spirit which is in him is tempered like iron, and made useful, instead of brittle and useless. But, if he carries on the softening and soothing process, in the next stage he begins to melt and waste, until he has wasted away his spirit and cut out the sinews of his soul; and he becomes a feeble warrior.

Very true.

If the element of spirit is naturally weak in him the change is speedily accomplished, but if he have a good deal, then the power of music weakening the spirit renders him excitable; – on the least provocation he flames up at once, and is speedily extinguished; instead of having spirit he grows irritable and passionate and is quite impracticable.

Exactly.

And so in gymnastics, if a man takes violent exercise and is a great feeder, and the reverse of a great student of music and philosophy, at first the high condition of his body fills him with pride and spirit, and he becomes twice the man that he was.

Certainly.

And what happens? If he do nothing else, and holds no converse with the Muses, does not even that intelligence which there may be in him, having no taste of any sort of learning or enquiry or thought or culture, grows feeble and dull and blind, his mind never waking up or receiving nourishment, and his senses not being purged of their mists?

True, he said.

And he ends by becoming a hater of philosophy, uncivilized, never using the weapon of persuasion, – he is like a wild beast, all violence and fierceness, and knows no other way of dealing; and he lives in all ignorance and evil conditions, and has no sense of propriety and grace.

That is quite true, he said.

And as there are two principles of human nature, one the spirited and the other the philosophical, some God, as I should say, has given mankind two arts answering to them (and only indirectly to the soul and body), in order that [412] these two principles ... may be relaxed or drawn tighter until they are duly harmonized.

That appears to be the intention.

And he who mingles music with gymnastic in the fairest proportions, and best attempers them to the soul, may be rightly called the true musician and harmonist in a far higher sense than the tuner of the strings.

You are quite right, Socrates.

And such a presiding genius will be always required in our State if the government is to last.

Yes, he will be absolutely necessary.

Such, then, are our principles of nurture and education: Where would be the use of going into further details about the dances of our citizens, or about their hunting and coursing, their gymnastic and equestrian contests? For these all follow the general principle, and having found that, we shall have no difficulty in discovering them.

I dare say that there will be no difficulty.

Very good, I said; then what is the next question? Must we not ask who are to be rulers and who subjects?

Certainly.

There can be no doubt that the elder must rule the younger.

Clearly.

And that the best of these must rule.

That is also clear.

Now, are not the best husbandmen those who are most devoted to husbandry?

Yes.

And as we are to have the best of guardians for our city, must they not be those who have most the character of guardians?

Yes.

And to this end they ought to be wise and efficient, and to have a special care of the State?

True.

And a man will be most likely to care about that which he loves?

To be sure.

And he will be most likely to love that which he regards as having the same interests with himself, and that of which the good or evil fortune is supposed by him at any time most to affect his own?

Very true, he replied.

Then there must be a selection. Let us note among the guardians those who in their whole life show the greatest eagerness to do what is for the good of their country, and the greatest repugnance to do what is against her interests.

Those are the right men.

And they will have to be watched at every age, in order that we may see whether they preserve their resolution, and never, under the influence either of force or enchantment, forget or cast off their sense of duty to the State.

How cast off? he said.

I will explain to you, I replied. A resolution may go out of a man's mind [413] either with his will or against his will; with his will when he gets rid of a falsehood and learns better, against his will whenever he is deprived of a truth.

I understand, he said, the willing loss of a resolution; the meaning of the unwilling I have yet to learn.

Why, I said, do you not see that men are unwillingly deprived of good, and willingly of evil? Is not to have lost the truth an evil, and to possess the truth a good? and you would agree that to conceive things as they are is to possess the truth?

Yes, he replied; I agree with you in thinking that mankind are deprived of truth against their will.

And is not this involuntary deprivation caused either by theft, or force, or enchantment?

Still, he replied, I do not understand you.

I fear that I must have been talking darkly, like the tragedians. I only mean that some men are changed by persuasion and that others forget; argument steals away the hearts of one class, and time of the other; and this I call theft. Now you understand me?

Yes.

Those again who are forced, are those whom the violence of some pain or grief compels to change their opinion.

I understand, he said, and you are quite right.

And you would also acknowledge that the enchanted are those who change their minds either under the softer influence of pleasure, or the sterner influence of fear?

Yes, he said; everything that deceives may be said to enchant.

Therefore, as I was just now saying, we must enquire who are the best guardians of their own conviction that what they think the interest of the State is to be the rule of their lives. We must watch them from their youth upwards, and make them perform actions in which they are most likely to forget or to be deceived, and he who remembers and is not deceived is to be selected, and he who fails in the trial is to be rejected. That will be the way?

Yes.

And there should also be toils and pains and conflicts prescribed for them, in which they will be made to give further proof of the same qualities.

Very right, he replied.

And then, I said, we must try them with enchantments – that is the third sort of test – and see what will be their behaviour: like those who take colts amid noise and tumult to see if they are of a timid nature, so must we take our youth amid terrors of some kind, and again pass them into pleasures, and prove them more thoroughly than gold is proved in the furnace, that we may discover whether they are armed against all enchantments, and of a noble bearing always, good guardians of themselves and of the music which they have learned, and retaining under all circumstances a rhythmical and harmonious nature, such as will be most serviceable to the individual and to the State. And he who at every age, as boy and youth and in mature life, has come out of the trial [414] victorious and pure, shall be appointed a ruler and guardian of the State; he shall be honoured in life and death, and shall receive sepulture and other memorials of honour, the greatest that we have to give. But him who fails, we must reject. I am inclined to think that this is the sort of way in which our rulers and guardians should be chosen and appointed. I speak generally, and not with any pretension to exactness.

And, speaking generally, I agree with you, he said.

And perhaps the word 'guardian' in the fullest sense ought to be applied to this higher class only, who preserve us against foreign enemies and maintain peace among our citizens at home, that the one may not have the will, or the others the power, to harm us. The young men whom we before called guardians may be more properly designated auxiliaries and supporters of the principles of the rulers.

I agree with you, he said.

How then may we devise one of those needful falsehoods of which we lately spoke – just one royal lie which may deceive the rulers, if that be possible, and at any rate the rest of the city?

What sort of lie? he said.

Nothing new, I replied; only an old Phoenician tale of what has often occurred before now in other places, (as the poets say, and have made the world believe) though not in our time, and I do not know whether such an event could ever happen again, or could now even be made probable, if it did.

How your words seem to hesitate on your lips!

You will not wonder, I replied, at my hesitation when you have heard.

Speak, he said, and fear not.

Well then, I will speak, although I really know not how to look you in the face, or in what words to utter the audacious fiction, which I propose to communicate gradually, first to the rulers, then to the soldiers, and lastly to the people. They are to be told that their youth was a dream, and the education and training which they received from us, an appearance only; in reality during all that time they were being formed and fed in the womb of the earth, where they themselves and their arms and appurtenances were manufactured; when they were completed, the earth, their mother, sent them up; and so, their country being their mother and also their nurse, they are bound to advise for her good, and to defend her against attacks, and her citizens they are to regard as children of the earth and their own brothers.

You had good reason, he said, to be ashamed of the lie which you were going to tell.

[415] True, I replied, but there is more coming; I have only told you half. Citizens, we shall say to them in our tale, you are brothers, yet God has framed you differently. Some of you have the power of command, and in the composition of these he has mingled gold, wherefore also they have the greatest honour; others he has made of silver, to be auxiliaries; others again who are to be husbandmen and craftsmen he has composed of brass and iron; and the species will generally be preserved in the children. But as all are of the same original stock, a golden parent will sometimes have a silver son, or a silver parent a golden son. And God proclaims as a first principle to the rulers, and above all else, that there is nothing which they should so anxiously guard, or of which they are to be such good guardians, as of the purity of the race. They should observe what elements mingle in their offspring; for if the son of a golden or silver parent has an

admixture of brass and iron, then nature orders a transposition of ranks, and the eye of the ruler must not be pitiful towards the child because he has to descend in the scale and become a husbandman or artisan, just as there may be sons of artisans who, having an admixture of gold or silver in them, are raised to honour, and become guardians or auxiliaries. For an oracle says that when a man of brass or iron guards the State, it will be destroyed. Such is the tale; is there any possibility of making our citizens believe in it?

Not in the present generation, he replied; there is no way of accomplishing this; but their sons may be made to believe in the tale, and their sons' sons, and posterity after them.

I see the difficulty, I replied; yet the fostering of such a belief will make them care more for the city and for one another. Enough, however, of the fiction, which may now fly abroad upon the wings of rumour, while we arm our earth-born heroes, and lead them forth under the command of their rulers. Let them look round and select a spot whence they can best suppress insurrection, if any prove refractory within, and also defend themselves against enemies, who like wolves may come down on the fold from without; there let them encamp, and when they have encamped, let them sacrifice to the proper Gods and prepare their dwellings.

Just so, he said.

And their dwellings must be such as will shield them against the cold of winter and the heat of summer.

I suppose that you mean houses, he replied.

Yes, I said; but they must be the houses of soldiers, and not of shop-keepers.

What is the difference? he said.

[416] That I will endeavour to explain, I replied. To keep watch-dogs, who, from want of discipline or hunger, or some evil habit or other, would turn upon the sheep and worry them, and behave not like dogs but wolves, would be a foul and monstrous thing in a shepherd?

Truly monstrous, he said.

And therefore every care must be taken that our auxiliaries, being stronger than our citizens, may not grow to be too much for them and become savage tyrants instead of friends and allies?

Yes, great care should be taken.

And would not a really good education furnish the best safeguard?

But they are well-educated already, he replied.

I cannot be so confident, my dear Glaucon, I said; I am much more certain that they ought to be, and that true education, whatever that may be, will have the greatest tendency to civilize and humanize them in their relations to one another, and to those who are under their protection.

Very true, he replied.

And not only their education, but their habitations, and all that belongs to them, should be such as will neither impair their virtue as guardians, nor tempt them to prey upon the other citizens. Any man of sense must acknowledge that.

He must.

Then now let us consider what will be their way of life, if they are to realize our idea of them. In the first place, none of them should have any property of his own beyond what is absolutely necessary; neither should they have a private house or store closed against anyone who has a mind to enter; their provisions should be only such as are required by trained warriors, who are men of temperance and courage; they should agree to receive from the citizens a fixed rate of pay, enough to meet the expenses of the year and no more; and they will go to mess and live together like soldiers in a camp. Gold and silver we will tell them that they have from God; the diviner metal is within them, and they have therefore no need of the dross which is current among men, and ought not to pollute the divine by any such earthly admixture; [417] for that commoner metal has been the source of many unholy deeds, but their own is undefiled. And they alone of all the citizens may not touch or handle silver or gold, or be under the same roof with them, or wear them, or drink from them. And this will be their salvation, and they will be the saviours of the State. But should they ever acquire homes or lands or moneys of their own, they will become housekeepers and husbandmen instead of guardians, enemies and tyrants instead of allies of the other citizens; hating and being hated, plotting and being plotted against, they will pass their whole life in much greater terror of internal than of external enemies, and the hour of ruin, both to themselves and to the rest of the State, will be at hand. For all which reasons may we not say that thus shall our State be ordered, and that these shall be the regulations appointed by us for guardians concerning their houses and all other matters?

Yes, said Glaucon.

BOOK IV

[419]

Here Adeimantus interposed a question: How would you answer, Socrates, said he, if a person were to say that you are making these people miserable, and that they are the cause of their own unhappiness; the city in fact belongs to them, but they are none the better for it; whereas other men acquire lands, and build large and handsome houses, and have everything handsome about them, offering sacrifices to the gods on their own account, and practising hospitality; moreover, as you were saying just now, they have gold and silver, and all that is usual among the favourites of fortune; but our poor citizens are no better than mercenaries [420] who are quartered in the city and are always mounting guard?

Yes, I said; and you may add that they are only fed, and not paid in addition to their food, like other men; and therefore they cannot, if they would, take a journey of pleasure; they have no money to spend on a mistress or any other luxurious fancy, which, as the world goes, is thought to be happiness; and many other accusations of the same nature might be added.

But, said he, let us suppose all this to be included in the charge.

You mean to ask, I said, what will be our answer?

Yes.

If we proceed along the old path, my belief, I said, is that we shall find the answer. And our answer will be that, even as they are, our guardians may very likely be the happiest of men; but that our aim in founding the State was not the disproportionate happiness of any one class, but the greatest happiness of the whole; we thought that in a State which is ordered with a view to the good of the whole we should be most likely to find justice, and in the ill-ordered State injustice: and, having found them, we might then decide which of the two is the happier. At present, I take it, we are fashioning the happy State, not piecemeal, or with a view of making a few happy citizens, but as a whole; and by-and-by we will proceed to view the opposite kind of State. Suppose that we were painting a statue, and someone came up to us and said, Why do you not put the most beautiful colours on the most beautiful parts of the body – the eyes ought to be purple, but you have made them black – to him we might fairly answer, Sir, you would not surely have us beautify the eyes to such a degree that they are no longer eyes; consider rather whether, by giving this and the other features their due proportion, we make the whole beautiful. And so I say to you, do not compel us to assign to the guardians a sort of happiness which will make them anything but guardians; for we too can clothe our husbandmen in royal apparel, and set crowns of gold on their heads, and bid them till the ground as much as they like, and no more. Our potters also might be allowed to repose on couches, and feast by the fireside, passing round the winecup, while their wheel is conveniently at hand, and working at pottery only as much as they like; in this way we might make every class happy – and then, as you imagine, the whole State would be happy. But do not put this idea into our heads; for, if [421] we listen to you, the husbandman will be no longer a husbandman, the potter will cease to be a potter, and no one will have the character of any distinct class in the State. Now this is not of much consequence where the corruption of society, and pretension to be what you are not, is confined to cobblers; but when the guardians of the laws and of the government are only seeming and not real guardians, then see how they turn the State upside down; and on the other hand they alone have the power of giving order and happiness to the State. We mean our guardians to be true saviours and not the destroyers of the State, whereas our opponent is thinking of peasants at a festival, who are enjoying a life of revelry, not of citizens who are doing their duty to the State. But, if so, we mean different things, and he is speaking of something which is not a State. And therefore we must consider whether in appointing our guardians we would look to their greatest happiness individually, or whether this principle of happiness does not rather reside in the State as a whole. But if the latter be the truth, then the guardians and auxiliaries, and all others equally with them, must be compelled or induced to do their own work in the best way. And thus the whole State will grow up in a noble order, and the several classes will receive the proportion of happiness which nature assigns to them.

I think that you are quite right.

I wonder whether you will agree with another remark which occurs to me.

What may that be?

There seem to be two causes of the deterioration of the arts.

What are they?

Wealth, I said, and poverty.

How do they act?

The process is as follows: When a potter becomes rich, will he, think you, any longer take the same pains with his art?

Certainly not.

He will grow more and more indolent and careless?

Very true.

And the result will be that he becomes a worse potter?

Yes; he greatly deteriorates.

But, on the other hand, if he has no money, and cannot provide himself with tools or instruments, he will not work equally well himself, nor will he teach his sons or apprentices to work equally well.

Certainly not.

Then, under the influence either of poverty or of wealth, workmen and their work are equally liable to degenerate?

That is evident.

Here, then, is a discovery of new evils, I said, against which the guardians will have to watch, or they will creep into the city unobserved.

What evils?

[422] Wealth, I said, and poverty; the one is the parent of luxury and indolence, and the other of meanness and viciousness, and both of discontent.

That is very true, he replied; but still I should like to know, Socrates, how our city will be able to go to war, especially against an enemy who is rich and powerful, if deprived of the sinews of war.

There would certainly be a difficulty, I replied, in going to war with one such enemy; but there is no difficulty where there are two of them.

How so? he asked.

In the first place, I said, if we have to fight, our side will be trained warriors fighting against an army of rich men.

That is true, he said.

And do you not suppose, Adeimantus, that a single boxer who was perfect in his art would easily be a match for two stout and well-to-do gentlemen who were not boxers?

Hardly, if they came upon him at once.

What, now, I said, if he were able to run away and then turn and strike at the one who first came up? And supposing he were to do this several times under the heat of a scorching sun, might he not, being an expert, overturn more than one stout personage?

Certainly, he said, there would be nothing wonderful in that.

And yet rich men probably have a greater superiority in the science and practise of boxing than they have in military qualities.

Likely enough.

Then we may assume that our athletes will be able to fight with two or three times their own number?

I agree with you, for I think you right.

And suppose that, before engaging, our citizens send an embassy to one of the two cities, telling them what is the truth: Silver and gold we neither have nor are permitted to have, but you may; do you therefore come and help us in war, and take the spoils of the other city: Who, on hearing these words, would choose to fight against lean wiry dogs, rather than, with the dogs on their side, against fat and tender sheep?

That is not likely; and yet there might be a danger to the poor State if the wealth of many States were to be gathered into one.

But how simple of you to use the term State at all of any but our own!

Why so?

You ought to speak of other States in the plural number; not one of them is a city, but many cities, as they say in the game. For indeed any city, however small, is in fact divided into two, [423] one the city of the poor, the other of the rich; these are at war with one another; and in either there are many smaller divisions, and you would be altogether beside the mark if you treated them all as a single State. But if you deal with them as many, and give the wealth or power or persons of the one to the others, you will always have a great many friends and not many enemies. And your State, while the wise order which has now been prescribed continues to prevail in her, will be the greatest of States, I do not mean to say in reputation or appearance, but in deed and truth, though she number not more than a thousand defenders. A single State which is her equal you will hardly find, either among Hellenes or barbarians, though many that appear to be as great and many times greater.

That is most true, he said.

And what, I said, will be the best limit for our rulers to fix when they are considering the size of the State and the amount of territory which they are to include, and beyond which they will not go?

What limit would you propose?

I would allow the State to increase so far as is consistent with unity; that, I think, is the proper limit.

Very good, he said.

Here then, I said, is another order which will have to be conveyed to our guardians: Let our city be accounted neither large nor small, but one and self-sufficing.

And surely, said he, this is not a very severe order which we impose upon them.

And the other, said I, of which we were speaking before is lighter still, – I mean the duty of degrading the offspring of the guardians when inferior, and of elevating into the rank of guardians the offspring of the lower classes, when naturally superior. The intention was, that, in the case of the citizens generally, each individual should be put to the use for which nature intended him, one-to-one work, and then every man would do his own business, and be one and not many; and so the whole city would be one and not many.

Yes, he said; that is not so difficult.

The regulations which we are prescribing, my good Adeimantus, are not, as might be supposed, a number of great principles, but trifles all, if care be taken, as the saying is, of the one great thing, – a thing, however, which I would rather call, not great, but sufficient for our purpose.

What may that be? he asked.

Education, I said, and nurture: If our citizens are well educated, and grow into sensible men, they will easily see their way through all these, as well as other matters which I omit; such, for example, [424] as marriage, the possession of women and the procreation of children, which will all follow the general principle that friends have all things in common, as the proverb says.

That will be the best way of settling them.

Also, I said, the State, if once started well, moves with accumulating force like a wheel. For good nurture and education implant good constitutions, and these good constitutions taking root in a good education improve more and more, and this improvement affects the breed in man as in other animals.

Very possibly, he said.

Then to sum up: This is the point to which, above all, the attention of our rulers should be directed, – that music and gymnastic be preserved in their original form, and no innovation made. They must do their utmost to maintain them intact. And when anyone says that mankind most regards,

'The newest song which the singers have,'

They will be afraid that he may be praising, not new songs, but a new kind of song; and this ought not to be praised, or conceived to be the meaning of the poet; for any musical innovation is full of danger to the whole State, and ought to be prohibited. So Damon tells me, and I can quite believe him; – he says that when modes of music change, the fundamental laws of the State always change with them.

Yes, said Adeimantus; and you may add my suffrage to Damon's and your own.

Then, I said, our guardians must lay the foundations of their fortress in music?

Yes, he said; the lawlessness of which you speak too easily steals in.

Yes, I replied, in the form of amusement; and at first sight it appears harmless.

Why, yes, he said, and there is no harm; were it not that little by little this spirit of licence, finding a home, imperceptibly penetrates into manners and customs; whence, issuing with greater force, it invades contracts between man and man, and from contracts goes on to laws and constitutions, in utter recklessness, ending at last, Socrates, by an overthrow of all rights, private as well as public.

Is that true? I said.

That is my belief, he replied.

Then, as I was saying, our youth should be trained from the first in a stricter system, for if amusements become lawless, and the youths themselves become lawless, they [425] can never grow up into well-conducted and virtuous citizens.

Very true, he said.

And when they have made a good beginning in play, and by the help of music have gained the habit of good order, then this habit of order, in a manner how unlike the lawless play of the others! will accompany them in all their actions and be a principle of growth to them, and if there be any fallen places in the State will raise them up again.

[minor omission, 425a-c]

Well, and about the business of the agora, and the ordinary dealings between man and man, or again about agreements with artisans; about insult and injury, or the commencement of actions, and the appointment of juries, what would you say? There may also arise questions about any impositions and exactions of market and harbour dues which may be required, and in general about the regulations of markets, police, harbours, and the like. But, oh heavens! shall we condescend to legislate on any of these particulars?

I think, he said, that there is no need to impose laws about them on good men; what regulations are necessary they will find out soon enough for themselves.

Yes, I said, my friend, if God will only preserve to them the laws which we have given them.

And without divine help, said Adeimantus, they will go on forever making and mending their laws and their lives in the hope of attaining perfection.

[425e-427a omitted]

[427a] I conceive, I said, that the true legislator will not trouble himself with this class of enactments whether concerning laws or the constitution either in an ill-ordered or in a well-ordered State; for in the former they are quite useless, and in the latter there will be no difficulty in devising them; and many of them will naturally flow out of our previous regulations.

What, then, he said, is still remaining to us of the work of legislation?

Nothing to us, I replied; but to Apollo, the God of Delphi, there remains the ordering of the greatest and noblest and chiefest things of all.

Which are they? he said.

The institution of temples and sacrifices, and the entire service of gods, demigods, and heroes; also the ordering of the repositories of the dead, and the rites which have to be observed by him who would propitiate the inhabitants of the world below. These are matters of which we are ignorant ourselves, and as founders of a city we should be unwise in trusting them to any interpreter but our ancestral deity. He is the god who sits in the centre, on the navel of the earth, and he is the interpreter of religion to all mankind.

You are right, and we will do as you propose.

But where, amid all this, is justice? Son of Ariston, tell me where. Now that our city has been made habitable, light a candle and search, and get your brother and Polemarchus and the rest of our friends to help, and let us see where in it we can discover justice and where injustice, and in what they differ from one another, and which of them the man who would be happy should have for his portion, whether seen or unseen by gods and men.

Nonsense, said Glaucon: did you not promise to search yourself, saying that for you not to help justice in her need would be an impiety?

I do not deny that I said so, and as you remind me, I will be as good as my word; but you must join.

We will, he replied.

Well, then, I hope to make the discovery in this way: I mean to begin with the assumption that our State, if rightly ordered, is perfect.

That is most certain.

And being perfect, is therefore wise and valiant and temperate and just.

That is likewise clear.

And whichever of these qualities we find in the State, [428] the one which is not found will be the residue?

Very good.

If there were four things, and we were searching for one of them, wherever it might be, the one sought for might be known to us from the first, and there would be no further trouble; or we might know the other three first, and then the fourth would clearly be the one left.

Very true, he said.

And is not a similar method to be pursued about the virtues, which are also four in number?

Clearly.

First among the virtues found in the State, wisdom comes into view, and in this I detect a certain peculiarity.

What is that?

The State which we have been describing is said to be wise as being good in counsel?

Very true.

And good counsel is clearly a kind of knowledge, for not by ignorance, but by knowledge, do men counsel well?

Clearly.

And the kinds of knowledge in a State are many and diverse?

Of course.

There is the knowledge of the carpenter; but is that the sort of knowledge which gives a city the title of wise and good in counsel?

Certainly not; that would only give a city the reputation of skill in carpentering.

Then a city is not to be called wise because possessing a knowledge which counsels for the best about wooden implements?

Certainly not.

Nor by reason of a knowledge which advises about brazen pots, I said, nor as possessing any other similar knowledge?

Not by reason of any of them, he said.

Nor yet by reason of a knowledge which cultivates the earth; that would give the city the name of agricultural?

Yes.

Well, I said, and is there any knowledge in our recently-founded State among any of the citizens which advises, not about any particular thing in the State, but about the whole, and considers how a State can best deal with itself and with other States?

There certainly is.

And what is this knowledge, and among whom is it found? I asked.

It is the knowledge of the guardians, he replied, and is found among those whom we were just now describing as perfect guardians.

And what is the name which the city derives from the possession of this sort of knowledge?

The name of good in counsel and truly wise.

And will there be in our city more of these true guardians or more smiths?

The smiths, he replied, will be far more numerous.

Will not the guardians be the smallest of all the classes who receive a name from the profession of some kind of knowledge?

Much the smallest.

And so by reason of the smallest part or class, and of the knowledge which resides in this presiding and ruling part of itself, the whole State, being thus constituted according to nature, will be wise; and this, [429] which has the only knowledge worthy to be called wisdom, has been ordained by nature to be of all classes the least.

Most true.

Thus, then, I said, the nature and place in the State of one of the four virtues has somehow or other been discovered.

And, in my humble opinion, very satisfactorily discovered, he replied.

Again, I said, there is no difficulty in seeing the nature of courage, and in what part that quality resides which gives the name of courageous to the State.

How do you mean?

Why, I said, everyone who calls any State courageous or cowardly, will be thinking of the part which fights and goes out to war on the State's behalf.

No one, he replied, would ever think of any other.

The rest of the citizens may be courageous or may be cowardly, but their courage or cowardice will not, as I conceive, have the effect of making the city either the one or the other.

Certainly not.

The city will be courageous in virtue of a portion of herself which preserves under all circumstances that opinion about the nature of things to be feared and not to be feared in which our legislator educated them; and this is what you term courage.

I should like to hear what you are saying once more, for I do not think that I perfectly understand you.

I mean that courage is a kind of salvation.

Salvation of what?

Of the opinion respecting things to be feared, what they are and of what nature, which the law implants through education; and I mean by the words 'under all circumstances' to intimate that in pleasure or in pain, or under the influence of desire or fear, a man preserves, and does not lose this opinion. Shall I give you an illustration?

If you please.

You know, I said, that dyers, when they want to dye wool for making the true sea-purple, begin by selecting their white colour first; this they prepare and dress with much care and pains, in order that the white ground may take the purple hue in full perfection. The dyeing then proceeds; and whatever is dyed in this manner becomes a fast colour, and no washing either with lyes or without them can take away the bloom. But, when the ground has not been duly prepared, you will have noticed how poor is the look either of purple or of any other colour.

Yes, he said; I know that they have a washed-out and ridiculous appearance.

Then now, I said, you will understand what our object was in selecting our soldiers, and [430] educating them in music and gymnastic; we were contriving influences which would prepare them to take the dye of the laws in perfection, and the colour of their opinion about dangers and of every other opinion was to be indelibly fixed by their nurture and training, not to be washed away by such potent lyes as pleasure – mightier agent far in washing the soul than any soda or lye; or by sorrow, fear, and desire, the mightiest of all other solvents. And this sort of universal saving power of true opinion in conformity with law about real and false dangers I call and maintain to be courage, unless you disagree.

But I agree, he replied; for I suppose that you mean to exclude mere uninstructed courage, such as that of a wild beast or of a slave – this, in your opinion, is not the courage which the law ordains, and ought to have another name.

Most certainly.

Then I may infer courage to be such as you describe?

Why, yes, said I, you may, and if you add the words 'of a citizen,' you will not be far wrong; – hereafter, if you like, we will carry the examination further, but at present we are seeking not for courage but justice; and for the purpose of our enquiry we have said enough.

You are right, he replied.

Two virtues remain to be discovered in the State – first, temperance, and then justice which is the end of our search.

Very true.

Now, can we find justice without troubling ourselves about temperance?

I do not know how that can be accomplished, he said, nor do I desire that justice should be brought to light and temperance lost sight of; and therefore I wish that you would do me the favour of considering temperance first.

Certainly, I replied, I should not be justified in refusing your request.

Then consider, he said.

Yes, I replied; I will; and as far as I can at present see, the virtue of temperance has more of the nature of harmony and symphony than the preceding.

How so? he asked.

Temperance, I replied, is the ordering or controlling of certain pleasures and desires; this is curiously enough implied in the saying of 'a man being his own master', and other traces of the same notion may be found in language.

No doubt, he said.

There is something ridiculous in the expression 'master of himself' for the master is also the servant and the servant [431] the master; and in all these modes of speaking the same person is denoted.

Certainly.

The meaning is, I believe, that in the human soul there is a better and also a worse principle; and when the better has the worse under control, then a man is said to be master of himself; and this is a term of praise: but when, owing to evil education or association, the better principle, which is also the smaller, is overwhelmed by the greater mass of the worse – in this case he is blamed and is called the slave of self and unprincipled.

Yes, there is reason in that.

And now, I said, look at our newly-created State, and there you will find one of these two conditions realized; for the State, as you will acknowledge, may be justly called master of itself, if the words 'temperance' and 'self-mastery' truly express the rule of the better part over the worse.

Yes, he said, I see that what you say is true.

Let me further note that the manifold and complex pleasures and desires and pains are generally found in children and women and servants, and in the freemen so called who are of the lowest and more numerous class.

Certainly, he said.

Whereas the simple and moderate desires which follow reason, and are under the guidance of mind and true opinion, are to be found only in a few, and those the best born and best educated.

Very true.

These two, as you may perceive, have a place in our State; and the meaner desires of the many are held down by the virtuous desires and wisdom of the few.

That I perceive, he said.

Then if there be any city which may be described as master of its own pleasures and desires, and master of itself, ours may claim such a designation?

Certainly, he replied.

It may also be called temperate, and for the same reasons?

Yes.

And if there be any State in which rulers and subjects will be agreed as to the question who are to rule, that again will be our State?

Undoubtedly.

And the citizens being thus agreed among themselves, in which class will temperance be found – in the rulers or in the subjects?

In both, as I should imagine, he replied.

[431e-432d, minor omissions]

And so, I said, we may consider three out of the four virtues to have been discovered in our State. The last of those qualities which make a state virtuous must be justice, if we only knew what that was.

The inference is obvious.

[...]

Truly, I said, we are stupid fellows.

Why so?

Why, my good sir, at the beginning of our enquiry, ages ago, there was justice tumbling out at our feet, and we never saw her; nothing could be more ridiculous. Like people who go about looking for what they have in their hands – that was the way with us – we looked not at what we were seeking, but at what was far off in the distance; and therefore, I suppose, we missed her.

What do you mean?

I mean to say that in reality for a long time past we have been talking of justice, and have failed to recognize her.

I grow impatient at the length of your exordium.

[433] Well then, tell me, I said, whether I am right or not: You remember the original principle which we were always laying down at the foundation of the State, that one man should practise one thing only, the thing to which his nature was best adapted; – now justice is this principle or a part of it.

Yes, we often said that one man should do one thing only.

Further, we affirmed that justice was doing one's own business, and not being a busybody; we said so again and again, and many others have said the same to us.

Yes, we said so.

Then to do one's own business in a certain way may be assumed to be justice. Can you tell me whence I derive this inference?

I cannot, but I should like to be told.

Because I think that this is the only virtue which remains in the State when the other virtues of temperance and courage and wisdom are abstracted; and, that this is the ultimate cause and condition of the existence of all of them, and while remaining in them is also their preservative; and we were saying that if the three were discovered by us, justice would be the fourth or remaining one.

That follows of necessity.

If we are asked to determine which of these four qualities by its presence contributes most to the excellence of the State, whether the agreement of rulers and subjects, or the preservation in the soldiers of the opinion which the law ordains about the true nature of dangers, or wisdom and watchfulness in the rulers, or whether this other which I am mentioning, and which is found in children and women, slave and freeman, artisan, ruler, subject, – the quality, I mean, of everyone doing his own work, and not being a busybody, would claim the palm – the question is not so easily answered.

Certainly, he replied, there would be a difficulty in saying which.

Then the power of each individual in the State to do his own work appears to compete with the other political virtues, wisdom, temperance, courage.

Yes, he said.

And the virtue which enters into this competition is justice?

Exactly.

Let us look at the question from another point of view: Are not the rulers in a State those to whom you would entrust the office of determining suits at law?

Certainly.

And are suits decided on any other ground but that a man may neither take what is another's, nor be deprived of what is his own?

Yes; that is their principle.

Which is a just principle?

Yes.

Then on this view also justice will be admitted to be the having and doing what is a man's own, and belongs to him?

[434] Very true.

Think, now, and say whether you agree with me or not. Suppose a carpenter to be doing the business of a cobbler, or a cobbler of a carpenter; and suppose them to exchange their implements or their duties, or the same person to be doing the work of both, or whatever be the change; do you think that any great harm would result to the State?

Not much.

But when the cobbler or any other man whom nature designed to be a trader, having his heart lifted up by wealth or strength or the number of his followers, or any like advantage, attempts to force his way into the class of warriors, or a warrior into that of legislators and guardians, for which he is unfitted, and either to take the implements or the duties of the other; or when one man is trader, legislator, and warrior all in one, then I think you will agree with me in saying that this interchange and this meddling of one with another is the ruin of the State.

Most true.

Seeing then, I said, that there are three distinct classes, any meddling of one with another, or the change of one into another, is the greatest harm to the State, and may be most justly termed evil-doing?

Precisely.

And the greatest degree of evil-doing to one's own city would be termed by you injustice?

Certainly.

This then is injustice; and on the other hand when the trader, the auxiliary, and the guardian each do their own business, that is justice, and will make the city just.

I agree with you.

We will not, I said, be over-positive as yet; but if, on trial, this conception of justice be verified in the individual as well as in the State, there will be no longer any room for doubt; if it be not verified, we must have a fresh enquiry. First let us complete the old investigation, which we began, as you remember, under the impression that, if we could previously examine justice on the larger scale, there would be less difficulty in discerning her in the individual. That larger example appeared to be the State, and accordingly we constructed as good a one as we could, knowing well that in the good State justice would be found. Let the discovery which we made be now applied to the individual – if they agree, we shall be satisfied; or, if there be a difference in the individual, [435] we will come back to the State and have another trial of the theory. The friction of the two when rubbed together may possibly strike a light in which justice will shine forth, and the vision which is then revealed we will fix in our souls.

That will be in regular course; let us do as you say.

I proceeded to ask: When two things, a greater and less, are called by the same name, are they like or unlike insofar as they are called the same?

Like, he replied.

The just man then, if we regard the idea of justice only, will be like the just State?

He will.

And a State was thought by us to be just when the three classes in the State severally did their own business; and also thought to be temperate and valiant and wise by reason of certain other affections and qualities of these same classes?

True, he said.

And so of the individual; we may assume that he has the same three principles in his own soul which are found in the State; and he may be rightly described in the same terms, because he is affected in the same manner?

Certainly, he said.

Once more then, O my friend, we have alighted upon an easy question – whether the soul has these three principles or not?

An easy question! Nay, rather, Socrates, the proverb holds that hard is the good.

Very true, I said; and I do not think that the method which we are employing is at all adequate to the accurate solution of this question; the true method is another and a longer one. Still we may arrive at a solution not below the level of the previous enquiry.

May we not be satisfied with that? he said; – under the circumstances, I am quite content.

I too, I replied, shall be extremely well satisfied.

Then faint not in pursuing the speculation, he said.

Must we not acknowledge, I said, that in each of us there are the same principles and habits which there are in the State; and that from the individual they pass into the State? – how else can they come there? Take the quality of passion or spirit; – it would be ridiculous to imagine that this quality, when found in States, is not derived from the individuals who are supposed to possess it, e.g. the Thracians, Scythians, and in general the northern nations; and the same may be said of the love of knowledge, which is the special characteristic of our [436] part of the world, or of the love of money, which may, with equal truth, be attributed to the Phoenicians and Egyptians.

Exactly so, he said.

There is no difficulty in understanding this.

None whatever.

But the question is not quite so easy when we proceed to ask whether these principles are three or one; whether, that is to say, we learn with one part of our nature, are angry with another, and with a third part desire the satisfaction of our natural appetites; or whether the whole soul comes into play in each sort of action – to determine that is the difficulty.

Yes, he said; there lies the difficulty.

Then let us now try and determine whether they are the same or different.

How can we? he asked.

I replied as follows: The same thing clearly cannot act or be acted upon in the same part or in relation to the same thing at the same time, in contrary ways; and therefore whenever this contradiction occurs in things apparently the same, we know that they are really not the same, but different.

Good.

For example, I said, can the same thing be at rest and in motion at the same time in the same part?

Impossible.

Still, I said, let us have a more precise statement of terms, lest we should hereafter fall out by the way. Imagine the case of a man who is standing and also moving his hands and his head, and suppose a person to say that one and the same person is in motion and at rest at the same moment – to such a mode of speech we should object, and should rather say that one part of him is in motion while another is at rest.

Very true.

[minor omission, 436d-437a]

Well, I said, would you not allow that assent and dissent, desire and aversion, attraction and repulsion, are all of them opposites, whether they are regarded as active or passive (for that makes no difference in the fact of their opposition)?

Yes, he said, they are opposites.

Well, I said, and hunger and thirst, and the desires in general, and again willing and wishing, – all these you would refer to the classes already mentioned. You would say – would you not? – that the soul of him who desires is seeking after the object of his desire; or that he is drawing to himself the thing which he wishes to possess: or again, when a person wants anything to be given him, his mind, longing for the realization of his desire, intimates his wish to have it by a nod of assent, as if he had been asked a question?

Very true.

And what would you say of unwillingness and dislike and the absence of desire; should not these be referred to the opposite class of repulsion and rejection?

Certainly.

Admitting this to be true of desire generally, let us suppose a particular class of desires, and out of these we will select hunger and thirst, as they are termed, which are the most obvious of them?

Let us take that class, he said.

The object of one is food, and of the other drink?

Yes.

And here comes the point: is not thirst the desire which the soul has of drink, and of drink only; not of drink qualified by anything else; for example, warm or cold, or much or little, or, in a word, drink of any particular sort: but if the thirst be accompanied by heat, then the desire is of cold drink; or, if accompanied by cold, then of warm drink; or, if the thirst be excessive, then the drink which is desired will be excessive; or, if not great, the quantity of drink will also be small: but thirst pure and simple will desire drink pure and simple, which is the natural satisfaction of thirst, as food is of hunger?

Yes, he said; the simple desire is, as you say, in every case of the simple object, and the qualified desire of the qualified object.

[minor omission, 438a-e]

[439] Would you not say that thirst is one of these essentially relative terms, having clearly a relation –

Yes, thirst is relative to drink.

And a certain kind of thirst is relative to a certain kind of drink; but thirst taken alone is neither of much nor little, nor of good nor bad, nor of any particular kind of drink, but of drink only?

Certainly.

Then the soul of the thirsty one, insofar as he is thirsty, desires only drink; for this he yearns and tries to obtain it?

That is plain.

And if you suppose something which pulls a thirsty soul away from drink, that must be different from the thirsty principle which draws him like a beast to drink; for, as we were saying, the same thing cannot at the same time with the same part of itself act in contrary ways about the same.

Impossible.

No more than you can say that the hands of the archer push and pull the bow at the same time, but what you say is that one hand pushes and the other pulls.

Exactly so, he replied.

And might a man be thirsty, and yet unwilling to drink?

Yes, he said, it constantly happens.

And in such a case what is one to say? Would you not say that there was something in the soul bidding a man to drink, and something else forbidding him, which is other and stronger than the principle which bids him?

I should say so.

And the forbidding principle is derived from reason, and that which bids and attracts proceeds from passion and disease?

Clearly.

Then we may fairly assume that they are two, and that they differ from one another; the one with which a man reasons, we may call the rational principle of the soul, the other, with which he loves and hungers and thirsts and feels the flutterings of any other desire, may be termed the irrational or appetitive, the ally of sundry pleasures and satisfactions?

Yes, he said, we may fairly assume them to be different.

Then let us finally determine that there are two principles existing in the soul. And what of passion, or spirit? Is it a third, or akin to one of the preceding?

I should be inclined to say – akin to desire.

Well, I said, there is a story which I remember to have heard, and in which I put faith. The story is, that Leontius, the son of Aglaion, coming up one day from the Piraeus, under the north wall on the outside, observed some dead bodies lying on the ground at the place of execution. He felt a desire to see them, and also a dread and abhorrence of them; [440] for a time he struggled and covered his eyes, but at length the desire got the better of him; and forcing them open, he ran up to the dead bodies, saying, Look, ye wretches, take your fill of the fair sight.

I have heard the story myself, he said.

The moral of the tale is, that anger at times goes to war with desire, as though they were two distinct things.

Yes; that is the meaning, he said.

And are there not many other cases in which we observe that when a man's desires violently prevail over his reason, he reviles himself, and is angry at the violence within him, and that in this struggle, which is like the struggle of factions in a State, his spirit is on the side of his reason; – but for the passionate or spirited element to take part with the desires when reason decides that she should not be opposed, is a sort of thing which I believe that you never observed occurring in yourself, nor, as I should imagine, in anyone else?

Certainly not.

Suppose that a man thinks he has done a wrong to another, the nobler he is the less able is he to feel indignant at any suffering, such as hunger, or cold, or any other pain which the injured person may inflict upon him – these he deems to be just, and, as I say, his anger refuses to be excited by them.

True, he said.

But when he thinks that he is the sufferer of the wrong, then he boils and chafes, and is on the side of what he believes to be justice; and because he suffers hunger or cold or other pain he is only the more determined to persevere and conquer. His noble spirit will not be quelled until he either slays or is slain; or until he hears the voice of the shepherd, that is, reason, bidding his dog bark no more.

The illustration is perfect, he replied; and in our State, as we were saying, the auxiliaries were to be dogs, and to hear the voice of the rulers, who are their shepherds.

I perceive, I said, that you quite understand me; there is, however, a further point which I wish you to consider.

What point?

You remember that passion or spirit appeared at first sight to be a kind of desire, but now we should say quite the contrary; for in the conflict of the soul spirit is arrayed on the side of the rational principle.

Most assuredly.

But a further question arises: Is passion different from reason also, or only a kind of reason; in which latter case, instead of three principles in the soul, there will only be two, the rational and the concupiscent; or rather, as the State was composed of three [441] classes, traders, auxiliaries, counsellors, so may there not be in the individual soul a third element which is passion or spirit, and when not corrupted by bad education is the natural auxiliary of reason?

Yes, he said, there must be a third.

Yes, I replied, if passion, which has already been shown to be different from desire, turn out also to be different from reason.

But that is easily proved: – We may observe even in young children that they are full of spirit almost as soon as they are born, whereas some of them never seem to attain to the use of reason, and most of them late enough.

Excellent, I said, and you may see passion equally in brute animals, which is a further proof of the truth of what you are saying. And we may once more appeal to the words of Homer, which have been already quoted by us,

'He smote his breast, and thus rebuked his soul,'

For in this verse Homer has clearly supposed the power which reasons about the better and worse to be different from the unreasoning anger which is rebuked by it.

Very true, he said.

And so, after much tossing, we have reached land, and are fairly agreed that the same principles which exist in the State exist also in the individual, and that they are three in number.

Exactly.

Must we not then infer that the individual is wise in the same way, and in virtue of the same quality which makes the State wise?

Certainly.

Also that the same quality which constitutes courage in the State constitutes courage in the individual, and that both the State and the individual bear the same relation to all the other virtues?

Assuredly.

And the individual will be acknowledged by us to be just in the same way in which the State is just?

That follows, of course.

We cannot but remember that the justice of the State consisted in each of the three classes doing the work of its own class?

We are not very likely to have forgotten, he said.

We must recollect that the individual in whom the several qualities of his nature do their own work will be just, and will do his own work?

Yes, he said, we must remember that too.

And ought not the rational principle, which is wise, and has the care of the whole soul, to rule, and the passionate or spirited principle to be the subject and ally?

Certainly.

And, as we were saying, the united influence of music and gymnastic will bring them into accord, nerving and [442] sustaining the reason with noble words and lessons, and moderating and soothing and civilizing the wildness of passion by harmony and rhythm?

Quite true, he said.

And these two, thus nurtured and educated, and having learned truly to know their own functions, will rule over the concupiscent, which in each of us is the largest part of the soul and by nature most insatiable of gain; over this they will keep guard, lest, waxing great and strong with the fullness of bodily pleasures, as they are termed, the concupiscent soul, no longer confined to her own sphere, should attempt to enslave and rule those who are not her natural-born subjects, and overturn the whole life of man?

Very true, he said.

Both together will they not be the best defenders of the whole soul and the whole body against attacks from without; the one counselling, and the other fighting under his leader, and courageously executing his commands and counsels?

True.

And he is to be deemed courageous whose spirit retains in pleasure and in pain the commands of reason about what he ought or ought not to fear?

Right, he replied.

And him we call wise who has in him that little part which rules, and which proclaims these commands; that part too being supposed to have a knowledge of what is for the interest of each of the three parts and of the whole?

Assuredly.

And would you not say that he is temperate who has these same elements in friendly harmony, in whom the one ruling principle of reason, and the two subject ones of spirit and desire are equally agreed that reason ought to rule, and do not rebel?

Certainly, he said, that is the true account of temperance whether in the State or individual.

And surely, I said, we have explained again and again how and by virtue of what quality a man will be just.

That is very certain.

And is justice dimmer in the individual, and is her form different, or is she the same which we found her to be in the State?

There is no difference in my opinion, he said.

Because, if any doubt is still lingering in our minds, a few commonplace instances will satisfy us of the truth of what I am saying.

What sort of instances do you mean?

If the case is put to us, must we not admit that the just State, or the man who is trained in the principles of such a State, will be less likely than the unjust to make away with a deposit of gold or silver? [443] Would anyone deny this?

No one, he replied.

Will the just man or citizen ever be guilty of sacrilege or theft, or treachery either to his friends or to his country?

Never.

Neither will he ever break faith where there have been oaths or agreements?

Impossible.

No one will be less likely to commit adultery, or to dishonour his father and mother, or to fail in his religious duties?

No one.

And the reason is that each part of him is doing its own business, whether in ruling or being ruled?

Exactly so.

Are you satisfied then that the quality which makes such men and such states is justice, or do you hope to discover some other?

Not I, indeed.

Then our dream has been realized; and the suspicion which we entertained at the beginning of our work of construction, that some divine power must have conducted us to a primary form of justice, has now been verified?

Yes, certainly.

And the division of labour which required the carpenter and the shoemaker and the rest of the citizens to be doing each his own business, and not another's, was a shadow of justice, and for that reason it was of use?

Clearly.

But in reality justice was such as we were describing, being concerned however, not with the outward man, but with the inward, which is the true self and concernment of man: for the just man does not permit the several elements within him to interfere with one another, or any of them to do the work of others, – he sets in order his own inner life, and is his own master and his own law, and at peace with himself; and when he has bound together the three principles within him, which may be compared to the higher, lower, and middle notes of the scale, and the intermediate intervals – when he has bound all these together, and is no longer many, but has become one entirely temperate and perfectly adjusted nature, then he proceeds to act, if he has to act, whether in a matter of property, or in the treatment of the body, or in some affair of politics or private business; always thinking and calling that which preserves and co-operates with this harmonious condition, just and good action, and the knowledge which presides over it, wisdom, [444] and that which at any time impairs this condition, he will call unjust action, and the opinion which presides over it ignorance.

You have said the exact truth, Socrates.

Very good; and if we were to affirm that we had discovered the just man and the just State, and the nature of justice in each of them, we should not be telling a falsehood?

Most certainly not.

May we say so, then?

Let us say so.

And now, I said, injustice has to be considered.

Clearly.

Must not injustice be a strife which arises among the three principles – a meddlesomeness, and interference, and rising up of a part of the soul against the whole, an assertion of unlawful authority, which is made by a rebellious subject against a true prince, of whom he is the natural vassal, – what is all this confusion and delusion but injustice, and intemperance and cowardice and ignorance, and every form of vice?

Exactly so.

And if the nature of justice and injustice be known, then the meaning of acting unjustly and being unjust, or, again, of acting justly, will also be perfectly clear?

What do you mean? he said.

Why, I said, they are like disease and health; being in the soul just what disease and health are in the body.

How so? he said.

Why, I said, that which is healthy causes health, and that which is unhealthy causes disease.

Yes.

And just actions cause justice, and unjust actions cause injustice?

That is certain.

And the creation of health is the institution of a natural order and government of one by another in the parts of the body; and the creation of disease is the production of a state of things at variance with this natural order?

True.

And is not the creation of justice the institution of a natural order and government of one by another in the parts of the soul, and the creation of injustice the production of a state of things at variance with the natural order?

Exactly so, he said.

Then virtue is the health and beauty and well-being of the soul, and vice the disease and weakness and deformity of the same?

True.

And do not good practices lead to virtue, and evil practices to vice?

Assuredly.

Still our old question of the comparative advantage of justice and injustice has not been answered: [445] Which is the more profitable, to be just and act justly and practise virtue, whether seen or unseen of gods and men, or to be unjust and act unjustly, if only unpunished and unreformed?

In my judgment, Socrates, the question has now become ridiculous. We know that, when the bodily constitution is gone, life is no longer endurable, though pampered with all kinds of meats and drinks, and having all wealth and all power; and shall we be told that when the very essence of the vital principle is undermined and corrupted, life is still worth having to a man, if only he be allowed to do whatever he likes with the single exception that he is not to acquire justice and virtue, or to escape from injustice and vice; assuming them both to be such as we have described?

Yes, I said, the question is, as you say, ridiculous. Still, as we are near the spot at which we may see the truth in the clearest manner with our own eyes, let us not faint by the way.

Certainly not, he replied.

Come up hither, I said, and behold the various forms of vice, those of them, I mean, which are worth looking at.

I am following you, he replied: proceed.

I said, The argument seems to have reached a height from which, as from some tower of speculation, a man may look down and see that virtue is one, but that the forms of vice are innumerable; there being four special ones which are deserving of note.

What do you mean? he said.

I mean, I replied, that there appear to be as many forms of the soul as there are distinct forms of the State.

How many?

There are five of the State, and five of the soul, I said.

What are they?

The first, I said, is that which we have been describing, and which may be said to have two names, monarchy and aristocracy, accordingly as rule is exercised by one distinguished man or by many.

True, he replied.

But I regard the two names as describing one form only; for whether the government is in the hands of one or many, if the governors have been trained in the manner which we have supposed, the fundamental laws of the State will be maintained.

That is true, he replied.

BOOK V

[449]

Such is the good and true City or State, and the good and true man is of the same pattern; and if this is right every other is wrong; and the evil is one which affects not only the ordering of the State, but also the regulation of the individual soul, and is exhibited in four forms.

What are they? he said.

I was proceeding to tell the order in which the four evil forms appeared to me to succeed one another, when Polemarchus, who was sitting a little way off, just beyond Adeimantus, began to whisper to him: stretching forth his hand, he took hold of the upper part of his coat by the shoulder, and drew him towards him, leaning forward himself so as to be quite close and saying something in his ear, of which I only caught the words, 'Shall we let him off, or what shall we do?'

Certainly not, said Adeimantus, raising his voice.

Who is it, I said, whom you are refusing to let off?

You, he said.

I repeated, Why am I especially not to be let off?

Why, he said, we think that you are lazy, and mean to cheat us out of a whole chapter which is a very important part of the story; and you fancy that we shall not notice your airy way of proceeding; as if it were self-evident to everybody, that in the matter of women and children, 'friends have all things in common.'

And was I not right, Adeimantus?

Yes, he said; but what is right in this particular case, like everything else, requires to be explained; for community may be of many kinds. Please, therefore, to say what sort of community you mean. We have been long expecting that you would tell us something about the family life of your citizens – how they will bring children into the world, and rear them when they have arrived, and, in general, what is the nature of this community of women and children – for we are of opinion that the right or wrong management of such matters will have a great and paramount influence on the State for good or for evil. And now, since the question is still undetermined, and you are taking in hand another State, we have resolved, as you [450] heard, not to let you go until you give an account of all this.

To that resolution, said Glaucon, you may regard me as saying Agreed.

And without more ado, said Thrasymachus, you may consider us all to be equally agreed.

I said, You know not what you are doing in thus assailing me: What an argument are you raising about the State! Just as

I thought that I had finished, and was only too glad that I had laid this question to sleep, and was reflecting how fortunate I was in your acceptance of what I then said, you ask me to begin again at the very foundation, ignorant of what a hornet's nest of words you are stirring. Now I foresaw this gathering trouble, and avoided it.

For what purpose do you conceive that we have come here, said Thrasymachus, – to look for gold, or to hear discourse?

Yes, but discourse should have a limit.

Yes, Socrates, said Glaucon, and the whole of life is the only limit which wise men assign to the hearing of such discourses. But never mind about us; take heart yourself and answer the question in your own way: What sort of community of women and children is this which is to prevail among our guardians? and how shall we manage the period between birth and education, which seems to require the greatest care? Tell us how these things will be.

Yes, my simple friend, but the answer is the reverse of easy; many more doubts arise about this than about our previous conclusions. For the practicability of what is said may be doubted; and looked at in another point of view, whether the scheme, if ever so practicable, would be for the best, is also doubtful. Hence I feel a reluctance to approach the subject, lest our aspiration, my dear friend, should turn out to be a dream only.

Fear not, he replied, for your audience will not be hard upon you; they are not sceptical or hostile.

[minor omission, 450d-451c]

For men born and educated like our citizens, the only way, in my opinion, of arriving at a right conclusion about the possession and use of women and children is to follow the path on which we originally started, when we said that the men were to be the guardians and watchdogs of the herd.

True.

Let us further suppose the birth and education of our women to be subject to similar or nearly similar regulations; then we shall see whether the result accords with our design.

What do you mean?

What I mean may be put into the form of a question, I said: Are dogs divided into hes and shes, or do they both share equally in hunting and in keeping watch and in the other duties of dogs? Or do we entrust to the males the entire and exclusive care of the flocks, while we leave the females at home, under the idea that the bearing and suckling their puppies is labour enough for them?

No, he said, they share alike; the only difference between them is that the males are stronger and the females weaker.

But can you use different animals for the same purpose, unless they are bred and fed in the same way?

You cannot.

Then, if women are to have the same duties as men, they must have the same nurture and education?

[452] Yes.

The education which was assigned to the men was music and gymnastic.

Yes.

Then women must be taught music and gymnastic and also the art of war, which they must practise like the men?

That is the inference, I suppose.

I should rather expect, I said, that several of our proposals, if they are carried out, being unusual, may appear ridiculous.

No doubt of it.

Yes, and the most ridiculous thing of all will be the sight of women naked in the palaestra, exercising with the men, especially when they are no longer young; they certainly will not be a vision of beauty, any more than the enthusiastic old men who in spite of wrinkles and ugliness continue to frequent the gymnasia.

Yes, indeed, he said: according to present notions the proposal would be thought ridiculous.

[minor omission, 452b-e]

First, then, whether the question is to be put in jest or in earnest, let us come to an understanding about the nature of woman: Is she [453] capable of sharing either wholly or partially in the actions of men, or not at all? And is the art of war one of those arts in which she can or can not share? That will be the best way of commencing the enquiry, and will probably lead to the fairest conclusion.

That will be much the best way.

Shall we take the other side first and begin by arguing against ourselves; in this manner the adversary's position will not be undefended.

Why not? he said.

Then let us put a speech into the mouths of our opponents. They will say: 'Socrates and Glaucon, no adversary need convict you, for you yourselves, at the first foundation of the State, admitted the principle that everybody was to do the one work suited to his own nature.' And certainly, if I am not mistaken, such an admission was made by us. 'And do not the natures of men and women differ very much indeed?' And we shall reply: Of course they do. Then we shall be asked, 'Whether the tasks assigned to men and to women should not be different, and such as are agreeable to their different natures?' Certainly they should. 'But if so, have you not fallen into a serious inconsistency in saying that men and women, whose natures are so entirely different, ought to perform the same actions?' – What defence will you make for us, my good Sir, against anyone who offers these objections?

That is not an easy question to answer when asked suddenly; and I shall and I do beg of you to draw out the case on our side.

These are the objections, Glaucon, and there are many others of a like kind, which I foresaw long ago; they made me afraid and reluctant to take in hand any law about the possession and nurture of women and children.

By Zeus, he said, the problem to be solved is anything but easy.

Why yes, I said, but the fact is that when a man is out of his depth, whether he has fallen into a little swimming bath or into mid-ocean, he has to swim all the same.

Very true.

And must not we swim and try to reach the shore: we will hope that Arion's dolphin or some other miraculous help may save us?

I suppose so, he said.

Well then, let us see if any way of escape can be found. We acknowledged – did we not? – that different natures ought to have different pursuits, and that men's and women's natures are different. And now what are we saying? – that different natures ought to have the same pursuits – this is the inconsistency which is charged upon us.

Precisely.

[454] Verily, Glaucon, I said, glorious is the power of the art of contradiction!

Why do you say so?

Because I think that many a man falls into the practice against his will. When he thinks that he is reasoning he is really disputing, just because he cannot define and divide, and so know that of which he is speaking; and he will pursue a merely verbal opposition in the spirit of contention and not of fair discussion.

Yes, he replied, such is very often the case; but what has that to do with us and our argument?

A great deal; for there is certainly a danger of our getting unintentionally into a verbal opposition.

In what way?

Why we valiantly and pugnaciously insist upon the verbal truth, that different natures ought to have different pursuits, but we never considered at all what was the meaning of sameness or difference of nature, or why we distinguished them when we assigned different pursuits to different natures and the same to the same natures.

Why, no, he said, that was never considered by us.

I said: Suppose that by way of illustration we were to ask the question whether there is not an opposition in nature between bald men and hairy men; and if this is admitted by us, then, if bald men are cobblers, we should forbid the hairy men to be cobblers, and conversely?

That would be a jest, he said.

Yes, I said, a jest; and why? because we never meant when we constructed the State, that the opposition of natures should extend to every difference, but only to those differences which affected the pursuit in which the individual is engaged; we should have argued, for example, that a physician and one who is in mind a physician may be said to have the same nature.

True.

Whereas the physician and the carpenter have different natures?

Certainly.

And if, I said, the male and female sex appear to differ in their fitness for any art or pursuit, we should say that such pursuit or art ought to be assigned to one or the other of them; but if the difference consists only in women bearing and men begetting children, this does not amount to a proof that a woman differs from a man in respect of the sort of education she should receive; and we shall therefore continue to maintain that our guardians and their wives ought to have the same pursuits.

Very true, he said.

Next, we shall ask our opponent how, in reference to any of the pursuits or arts of civic life, the nature of a woman differs from that of a man?

[minor omission, 454e-455b]

Let us say to him: Come now, and we will ask you a question: – when you spoke of a nature gifted or not gifted in any respect, did you mean to say that one man will acquire a thing easily, another with difficulty; a little learning will lead the one to discover a great deal; whereas the other, after much study and application, no sooner learns than he forgets; or again, did you mean, that the one has a body which is a good servant to his mind, while the body of the other is a hindrance to him? – would not these be the sort of differences which distinguish the man gifted by nature from the one who is ungifted?

No one will deny that.

And can you mention any pursuit of mankind in which the male sex has not all these gifts and qualities in a higher degree than the female? Need I waste time in speaking of the art of weaving, and the management of pancakes and preserves, in which womankind does really appear to be great, and in which for her to be beaten by a man is of all things the most absurd?

You are quite right, he replied, in maintaining the general inferiority of the female sex: although many women are in many things superior to many men, yet on the whole what you say is true.

And if so, my friend, I said, there is no special faculty of administration in a state which a woman has because she is a woman, or which a man has by virtue of his sex, but the gifts of nature are alike diffused in both; all the pursuits of men are the pursuits of women also, but in all of them a woman is inferior to a man.

Very true.

Then are we to impose all our enactments on men and none of them on women?

That will never do.

One woman has a gift of healing, another not; one is a musician, and another has no music in her nature?

Very true.

[456] And one woman has a turn for gymnastic and military exercises, and another is unwarlike and hates gymnastics?

Certainly.

And one woman is a philosopher, and another is an enemy of philosophy; one has spirit, and another is without spirit?

That is also true.

Then one woman will have the temper of a guardian, and another not. Was not the selection of the male guardians determined by differences of this sort?

Yes.

Men and women alike possess the qualities which make a guardian; they differ only in their comparative strength or weakness.

Obviously.

And those women who have such qualities are to be selected as the companions and colleagues of men who have similar qualities and whom they resemble in capacity and in character?

Very true.

And ought not the same natures to have the same pursuits?

They ought.

Then, as we were saying before, there is nothing unnatural in assigning music and gymnastic to the wives of the guardians – to that point we come round again.

Certainly not.

The law which we then enacted was agreeable to nature, and therefore not an impossibility or mere aspiration; and the contrary practice, which prevails at present, is in reality a violation of nature.

That appears to be true.

We had to consider, first, whether our proposals were possible, and secondly whether they were the most beneficial?

Yes.

And the possibility has been acknowledged?

Yes.

The very great benefit has next to be established?

Quite so.

You will admit that the same education which makes a man a good guardian will make a woman a good guardian; for their original nature is the same?

Yes.

I should like to ask you a question.

What is it?

Would you say that all men are equal in excellence, or is one man better than another?

The latter.

And in the commonwealth which we were founding do you conceive the guardians who have been brought up on our model system to be more perfect men, or the cobblers whose education has been cobbling?

What a ridiculous question!

You have answered me, I replied: Well, and may we not further say that our guardians are the best of our citizens?

By far the best.

And will not their wives be the best women?

Yes, by far the best.

And can there be anything better for the interests of the State than that the men and women of a State should be as good as possible?

There can be nothing better.

And this is what the arts of music and gymnastic, when present in such manner as [457] we have described, will accomplish?

Certainly.

Then we have made an enactment not only possible but in the highest degree beneficial to the State?

True.

Then let the wives of our guardians strip, for their virtue will be their robe, and let them share in the toils of war and the defence of their country; only in the distribution of labours the lighter are to be assigned to the women, who are the weaker natures, but in other respects their duties are to be the same. And as for the man who laughs at naked women exercising their bodies from the best of motives, in his laughter he is plucking,

'A fruit of unripe wisdom,'

And he himself is ignorant of what he is laughing at, or what he is about; – for that is, and ever will be, the best of sayings, that the useful is the noble and the hurtful is the base.

Very true.

Here, then, is one difficulty in our law about women, which we may say that we have now escaped; the wave has not swallowed us up alive for enacting that the guardians of either sex should have all their pursuits in common; to the utility and also to the possibility of this arrangement the consistency of the argument with itself bears witness.

Yes, that was a mighty wave which you have escaped.

Yes, I said, but a greater is coming; you will not think much of this when you see the next.

Go on; let me see.

The law, I said, which is the sequel of this and of all that has preceded, is to the following effect, – 'that the wives of our guardians are to be common, and their children are to be common, and no parent is to know his own child, nor any child his parent.'

Yes, he said, that is a much greater wave than the other; and the possibility as well as the utility of such a law are far more questionable.

[minor omission, 457d-458b]

First of all, then, if you have no objection, I will endeavour with your help to consider the advantages of the measure; and hereafter the question of possibility.

I have no objection; proceed.

First, I think that if our rulers and their auxiliaries are to be worthy of the name which they bear, there must be willingness to obey in the one and the power of command in the other; the guardians must themselves obey the laws, and they must also imitate the spirit of them in any details which are entrusted to their care.

That is right, he said.

You, I said, who are their legislator, having selected the men, will now select the women and give them to them; – they must be as far as possible of like natures with them; and they must live in common houses and meet at common meals. None of them will have anything specially his or her own; they will be together, and will be brought up together, and will associate at gymnastic exercises. And so they will be drawn by a necessity of their natures to have intercourse with each other – necessity is not too strong a word, I think?

Yes, he said; – necessity, not geometrical, but another sort of necessity which lovers know, and which is far more convincing and constraining to the mass of mankind.

True, I said; and this, Glaucon, like all the rest, must proceed after an orderly fashion; in a city of the blessed, licentiousness is an unholy thing which the rulers will forbid.

Yes, he said, and it ought not to be permitted.

Then clearly the next thing will be to make matrimony sacred in the highest degree, and what is most beneficial will be deemed sacred?

Exactly.

[459] And how can marriages be made most beneficial? – that is a question which I put to you, because I see in your house dogs for hunting, and of the nobler sort of birds not a few. Now, I beseech you, do tell me, have you ever attended to their pairing and breeding?

In what particulars?

Why, in the first place, although they are all of a good sort, are not some better than others?

True.

And do you breed from them all indifferently, or do you take care to breed from the best only?

From the best.

And do you take the oldest or the youngest, or only those of ripe age?

I choose only those of ripe age.

And if care was not taken in the breeding, your dogs and birds would greatly deteriorate?

Certainly.

And the same of horses and animals in general?

Undoubtedly.

Good heavens! My dear friend, I said, what consummate skill will our rulers need if the same principle holds of the human species!

Certainly, the same principle holds; but why does this involve any particular skill?

Because, I said, our rulers will often have to practise upon the body corporate with medicines. Now you know that when patients do not require medicines, but have only to be put under a regimen, the inferior sort of practitioner is deemed to be good enough; but when medicine has to be given, then the doctor should be more of a man.

That is quite true, he said; but to what are you alluding?

I mean, I replied, that our rulers will find a considerable dose of falsehood and deceit necessary for the good of their subjects: we were saying that the use of all these things regarded as medicines might be of advantage.

And we were very right.

And this lawful use of them seems likely to be often needed in the regulations of marriages and births.

How so?

Why, I said, the principle has been already laid down that the best of either sex should be united with the best as often, and the inferior with the inferior, as seldom as possible; and that they should rear the offspring of the one sort of union, but not of the other, if the flock is to be maintained in first-rate condition. Now these goings on must be a secret which the rulers only know, or there will be a further danger of our herd, as the guardians may be termed, breaking out into rebellion.

Very true.

Had we not better appoint certain festivals at which we will bring together the brides and bridegrooms, and sacrifices will be offered and suitable hymeneal songs composed [460] by our poets: the number of weddings is a matter which must be left to the discretion of the rulers, whose aim will be to preserve the average of population? There are many other things which they will have to consider, such as the effects of wars and diseases and any similar agencies, in order as far as this is possible to prevent the State from becoming either too large or too small.

Certainly, he replied.

We shall have to invent some ingenious kind of lots which the less worthy may draw on each occasion of our bringing them together, and then they will accuse their own ill-luck and not the rulers.

To be sure, he said.

And I think that our braver and better youth, besides their other honours and rewards, might have greater facilities of intercourse with women given them; their bravery will be a reason, and such fathers ought to have as many sons as possible.

True.

And the proper officers, whether male or female or both, for offices are to be held by women as well as by men –

Yes –

The proper officers will take the offspring of the good parents to the pen or fold, and there they will deposit them with certain nurses who dwell in a separate quarter; but the offspring of the inferior, or of the better when they chance to be deformed, will be put away in some mysterious, unknown place, as they should be.

Yes, he said, that must be done if the breed of the guardians is to be kept pure.

They will provide for their nurture, and will bring the mothers to the fold when they are full of milk, taking the greatest possible care that no mother recognises her own child; and other wet-nurses may be engaged if more are required.

Care will also be taken that the process of suckling shall not be protracted too long; and the mothers will have no getting up at night or other trouble, but will hand over all this sort of thing to the nurses and attendants.

You suppose the wives of our guardians to have a fine, easy time of it when they are having children.

Why, said I, and so they ought. Let us, however, proceed with our scheme. We were saying that the parents should be in the prime of life?

Very true.

And what is the prime of life? May it not be defined as a period of about twenty years in a woman's life, and thirty in a man's?

Which years do you mean to include?

A woman, I said, at twenty years of age may begin to bear children to the State, and continue to bear them until forty; a man may begin at five-and-twenty, when he has passed the point at which the pulse of life beats quickest, and continue to beget children until he be fifty-five.

[461] Certainly, he said, both in men and women those years are the prime of physical as well as of intellectual vigour.

Anyone above or below the prescribed ages who takes part in the public hymeneals shall be said to have done an unholy and unrighteous thing; the child of which he is the father, if it steals into life, will have been conceived under auspices very unlike the sacrifices and prayers, which at each hymeneal priestesses and priest and the whole city will offer, that the new generation may be better and more useful than their good and useful parents, whereas his child will be the offspring of darkness and strange lust.

Very true, he replied.

And the same law will apply to any one of those within the prescribed age who forms a connection with any woman in the prime of life without the sanction of the rulers; for we shall say that he is raising up a bastard to the State, uncertified and unconsecrated.

Very true, he replied.

This applies, however, only to those who are within the specified age: after that we allow them to range at will, except that a man may not marry his daughter or his daughter's daughter, or his mother or his mother's mother; and women, on the other hand, are prohibited from marrying their sons or fathers, or son's son or father's father, and so on in either direction. And we grant all this, accompanying the permission with strict orders to prevent any embryo which may come into being from seeing the light; and if any force a way to the birth, the parents must understand that the offspring of such a union cannot be maintained, and arrange accordingly.

That also, he said, is a reasonable proposition. But how will they know who are fathers and daughters, and so on?

They will never know. The way will be this: – dating from the day of the hymeneal, the bridegroom who was then married will call all the male children who are born in the seventh and tenth month afterwards his sons, and the female children his daughters, and they will call him father, and he will call their children his grandchildren, and they will call the elder generation grandfathers and grandmothers. All who were begotten at the time when their fathers and mothers came together will be called their brothers and sisters, and these, as I was saying, will be forbidden to inter-marry. This, however, is not to be understood as an absolute prohibition of the marriage of brothers and sisters; if the lot favours them, and they receive the sanction of the Pythian oracle, the law will allow them.

Quite right, he replied.

Such is the scheme, Glaucon, according to which the guardians of our State are to have their wives and families in common. And now you would have the argument show that this community is consistent with the rest of our polity, and also that nothing can be better – would you not?

[462] Yes, certainly.

Shall we try to find a common basis by asking of ourselves what ought to be the chief aim of the legislator in making laws and in the organization of a State, – what is the greatest good, and what is the greatest evil, and then consider whether our previous description has the stamp of the good or of the evil?

By all means.

Can there be any greater evil than discord and distraction and plurality where unity ought to reign? Or any greater good than the bond of unity?

There cannot.

And there is unity where there is community of pleasures and pains – where all the citizens are glad or grieved on the same occasions of joy and sorrow?

No doubt.

Yes; and where there is no common but only private feeling a State is disorganized – when you have one half of the world triumphing and the other plunged in grief at the same events happening to the city or the citizens?

Certainly.

Such differences commonly originate in a disagreement about the use of the terms 'mine' and 'not mine,' 'his' and 'not his.'

Exactly so.

And is not that the best-ordered State in which the greatest number of persons apply the terms 'mine' and 'not mine' in the same way to the same thing?

Quite true.

Or that again which most nearly approaches to the condition of the individual – as in the body, when but a finger of one of us is hurt, the whole frame, drawn towards the soul as a centre and forming one kingdom under the ruling power therein, feels the hurt and sympathizes all together with the part affected, and we say that the man has a pain in his finger; and the same expression is used about any other part of the body, which has a sensation of pain at suffering or of pleasure at the alleviation of suffering.

Very true, he replied; and I agree with you that in the best-ordered State there is the nearest approach to this common feeling which you describe.

Then when any one of the citizens experiences any good or evil, the whole State will make his case their own, and will either rejoice or sorrow with him?

Yes, he said, that is what will happen in a well-ordered State.

It will now be time, I said, for us to return to our State and see whether this or some other form is most in accordance with these fundamental principles.

Very good.

[463] Our State like every other has rulers and subjects?

True.

All of whom will call one another citizens?

Of course.

But is there not another name which people give to their rulers in other States?

Generally, they call them masters, but in democratic States they simply call them rulers.

And in our State what other name besides that of citizens do the people give the rulers?

They are called saviours and helpers, he replied.

And what do the rulers call the people?

Their maintainers and foster-fathers.

And what do they call them in other States?

Slaves.

And what do the rulers call one another in other States?

Fellow-rulers.

And what in ours?

Fellow-guardians.

Did you ever know an example in any other State of a ruler who would speak of one of his colleagues as his friend and of another as not being his friend?

Yes, very often.

And the friend he regards and describes as one in whom he has an interest, and the other as a stranger in whom he has no interest?

Exactly.

But would any of your guardians think or speak of any other guardian as a stranger?

Certainly he would not; for everyone whom they meet will be regarded by them either as a brother or sister, or father or mother, or son or daughter, or as the child or parent of those who are thus connected with him.

Capital, I said; but let me ask you once more: Shall they be a family in name only; or shall they in all their actions be true to the name? For example, in the use of the word 'father,' would the care of a father be implied and the filial reverence and duty and obedience to him which the law commands; and is the violator of these duties to be regarded as an impious and unrighteous person who is not likely to receive much good either at the hands of God or of man? Are these to be or not to be the strains which the children will hear repeated in their ears by all the citizens about those who are intimated to them to be their parents and the rest of their kinsfolk?

These, he said, and none other; for what can be more ridiculous than for them to utter the names of family ties with the lips only and not to act in the spirit of them?

Then in our city the language of harmony and concord will be more often heard than in any other. As I was describing before, when anyone is well or ill, the universal word will be 'with me it is well' or 'it is ill.'

Most true.

[464] And agreeably to this mode of thinking and speaking, were we not saying that they will have their pleasures and pains in common?

Yes, and so they will.

And they will have a common interest in the same thing which they will alike call 'my own,' and having this common interest they will have a common feeling of pleasure and pain?

Yes, far more so than in other States.

And the reason of this, over and above the general constitution of the State, will be that the guardians will have a community of women and children?

That will be the chief reason.

And this unity of feeling we admitted to be the greatest good, as was implied in our own comparison of a well-ordered State to the relation of the body and the members, when affected by pleasure or pain?

That we acknowledged, and very rightly.

Then the community of wives and children among our citizens is clearly the source of the greatest good to the State?

Certainly.

And this agrees with the other principle which we were affirming, – that the guardians were not to have houses or lands or any other property; their pay was to be their food, which they were to receive from the other citizens, and they were to have no private expenses; for we intended them to preserve their true character of guardians.

Right, he replied.

Both the community of property and the community of families, as I am saying, tend to make them more truly guardians; they will not tear the city in pieces by differing about 'mine' and 'not mine;' each man dragging any acquisition which he has made into a separate house of his own, where he has a separate wife and children and private pleasures and pains; but all will be affected as far as may be by the same pleasures and pains because they are all of one opinion about what is near and dear to them, and therefore they all tend towards a common end.

Certainly, he replied.

And as they have nothing but their persons which they can call their own, suits and complaints will have no existence among them; they will be delivered from all those quarrels of which money or children or relations are the occasion.

Of course they will.

Neither will trials for assault or insult ever be likely to occur among them. For that equals should defend themselves against equals we shall maintain to be honourable and right; we shall make the protection of the person a matter of necessity.

That is good, he said.

[465] Yes; and there is a further good in the law; viz. that if a man has a quarrel with another he will satisfy his resentment then and there, and not proceed to more dangerous lengths.

Certainly.

To the elder shall be assigned the duty of ruling and chastising the younger.

Clearly.

Nor can there be a doubt that the younger will not strike or do any other violence to an elder, unless the magistrates command him; nor will he slight him in any way. For there are two guardians, shame and fear, mighty to prevent him: shame, which makes men refrain from laying hands on those who are to them in the relation of parents; fear, that the injured one will be succoured by the others who are his brothers, sons, fathers.

That is true, he replied.

Then in every way the laws will help the citizens to keep the peace with one another?

Yes, there will be no want of peace.

And as the guardians will never quarrel among themselves there will be no danger of the rest of the city being divided either against them or against one another.

None whatever.

[465c-466c omitted]

You agree then, I said, that men and women are to have a common way of life such as we have described – common education, common children; and they are to watch over the citizens in common whether abiding in the city or going out to war; they are to keep watch together, and to hunt together like dogs; and always and in all things, as far as they are able, women are to share with the men? And in so doing they will do what is best, and will not violate, but preserve the natural relation of the sexes.

I agree with you, he replied.

The enquiry, I said, has yet to be made, whether such a community be found possible – as among other animals, so also among men – and if possible, in what way possible?

You have anticipated the question which I was about to suggest.

There is no difficulty, I said, in seeing how war will be carried on by them.

How?

Why, of course they will go on expeditions together; and will take with them any of their children who are strong enough, that, after the manner of the artisan's child, they may look on at the work which they will have to do when they are grown up; [467] and besides looking on they will have to help and be of use in war, and to wait upon their fathers and mothers. Did you never observe in the arts how the potters' boys look on and help, long before they touch the wheel?

Yes, I have.

And shall potters be more careful in educating their children and in giving them the opportunity of seeing and practising their duties than our guardians will be?

The idea is ridiculous, he said.

There is also the effect on the parents, with whom, as with other animals, the presence of their young ones will be the greatest incentive to valour.

That is quite true, Socrates; and yet if they are defeated, which may often happen in war, how great the danger is! The children will be lost as well as their parents, and the State will never recover.

True, I said; but would you never allow them to run any risk?

I am far from saying that.

Well, but if they are ever to run a risk should they not do so on some occasion when, if they escape disaster, they will be the better for it?

Clearly.

[minor omission, 467b-468a]

Next, as to war; what are to be the relations of your soldiers to one another and to their enemies? I should be inclined to propose that the soldier who leaves his rank or throws away his arms, or is guilty of any other act of cowardice, should be degraded into the rank of a husbandman or artisan. What do you think?

By all means, I should say.

And he who allows himself to be taken prisoner may as well be made a present of to his enemies; he is their lawful prey, and let them do what they like with him.

Certainly.

But the hero who has distinguished himself, what shall be done to him? In the first place, he shall receive honour in the army from his youthful comrades; every one of them in succession shall crown him. What do you say?

I approve.

And what do you say to his receiving the right hand of fellowship?

To that too, I agree.

But you will hardly agree to my next proposal.

What is your proposal?

That he should kiss and be kissed by them.

Most certainly, and I should be disposed to go further, and say: Let no one whom he has a mind to kiss refuse to be kissed by him while the expedition lasts. So that if there be a lover in the army, whether his love be youth or maiden, he may be more eager to win the prize of valour.

Capital, I said. That the brave man is to have more wives than others has been already determined: and he is to have first choices in such matters more than others, in order that he may have as many children as possible?

Agreed.

[minor omission, 468c-e]

Yes, I said; and when a man dies gloriously in war shall we not say, in the first place, that he is of the golden race?

To be sure.

Nay, have we not the authority of Hesiod for affirming that when they are dead,

[469] 'They are holy angels upon the earth, authors of good, averters of evil, the guardians of speech-gifted men'?

Yes; and we accept his authority.

We must learn of the god how we are to order the sepulture of divine and heroic personages, and what is to be their special distinction; and we must do as he bids?

By all means.

And in ages to come we will reverence them and kneel before their sepulchres as at the graves of heroes. And not only they but any who are deemed pre-eminently good, whether they die from age, or in any other way, shall be admitted to the same honours.

That is very right, he said.

Next, how shall our soldiers treat their enemies? What about this?

In what respect do you mean?

First of all, in regard to slavery? Do you think it right that Hellenes should enslave Hellenic States, or allow others to enslave them, if they can help? Should not their custom be to spare them, considering the danger which there is that the whole race may one day fall under the yoke of the barbarians?

To spare them is infinitely better.

Then no Hellene should be owned by them as a slave; that is a rule which they will observe and advise the other Hellenes to observe.

Certainly, he said; they will in this way be united against the barbarians and will keep their hands off one another.

Next as to the slain; ought the conquerors, I said, to take anything but their armour? Does not the practice of despoiling an enemy afford an excuse for not facing the battle? Cowards skulk about the dead, pretending that they are fulfilling a duty, and many an army before now has been lost from this love of plunder.

Very true.

[minor omission, 469d-e]

Neither shall we offer up arms at the temples of the gods, least of all the arms of Hellenes, if we care to [470] maintain good feeling with other Hellenes; and, indeed, we have reason to fear that the offering of spoils taken from kinsmen may be a pollution unless commanded by the god himself?

Very true.

Again, as to the devastation of Hellenic territory or the burning of houses, what is to be the practice?

May I have the pleasure, he said, of hearing your opinion?

Both should be forbidden, in my judgment; I would take the annual produce and no more. Shall I tell you why?

Pray do.

Why, you see, there is a difference in the names 'discord' and 'war,' and I imagine that there is also a difference in their natures; the one is expressive of what is internal and domestic, the other of what is external and foreign; and the first of the two is termed discord, and only the second, war.

That is a very proper distinction, he replied.

And may I not observe with equal propriety that the Hellenic race is all united together by ties of blood and friendship, and alien and strange to the barbarians?

Very good, he said.

And therefore when Hellenes fight with barbarians and barbarians with Hellenes, they will be described by us as being at war when they fight, and by nature enemies, and this kind of antagonism should be called war; but when Hellenes fight with one another we shall say that Hellas is then in a state of disorder and discord, they being by nature friends; and such enmity is to be called discord.

I agree.

Consider then, I said, when that which we have acknowledged to be discord occurs, and a city is divided, if both parties destroy the lands and burn the houses of one another, how wicked does the strife appear! No true lover of his country would bring himself to tear in pieces his own nurse and mother: There might be reason in the conqueror depriving the conquered of their harvest, but still they would have the idea of peace in their hearts and would not mean to go on fighting forever.

Yes, he said, that is a better temper than the other.

And will not the city, which you are founding, be an Hellenic city?

It ought to be, he replied.

[minor omission, 470e-471a]

And as they are Hellenes themselves they will not devastate Hellas, nor will they burn houses, nor ever suppose that the whole population of a city – men, women, and children – are equally their enemies, for they know that the guilt of war is always confined to a few persons and that the many are their friends. And for all these reasons they will be unwilling to waste their lands and rase their houses; their enmity to them will only last until the many innocent sufferers have compelled the guilty few to give satisfaction?

I agree, he said, that our citizens should thus deal with their Hellenic enemies; and with barbarians as the Hellenes now deal with one another.

Then let us enact this law also for our guardians: – that they are neither to devastate the lands of Hellenes nor to burn their houses.

Agreed; and we may agree also in thinking that these, like all our previous enactments, are very good.

But still I must say, Socrates, that if you are allowed to go on in this way you will entirely forget the other question which at the commencement of this discussion you thrust aside: – Is such an order of things possible, and how, if at all?

[minor omission, 471c-e]

[472] If I loiter for a moment, you instantly make a raid upon me, I said, and have no mercy; I have hardly escaped the first and second waves, and you seem not to be aware that you are now bringing upon me the third, which is the greatest and heaviest. When you have seen and heard the third wave, I think you will be more considerate and will acknowledge that some fear and hesitation was natural respecting a proposal so extraordinary as that which I have now to state and investigate.

The more appeals of this sort which you make, he said, the more determined are we that you shall tell us how such a State is possible: speak out and at once.

Let me begin by reminding you that we found our way hither in the search after justice and injustice.

True, he replied; but what of that?

I was only going to ask whether, if we have discovered them, we are to require that the just man should in nothing fail of absolute justice; or may we be satisfied with an approximation, and the attainment in him of a higher degree of justice than is to be found in other men?

The approximation will be enough.

We were enquiring into the nature of absolute justice and into the character of the perfectly just, and into injustice and the perfectly unjust, that we might have an ideal. We were to look at these in order that we might judge of our own happiness and unhappiness according to the standard which they exhibited and the degree in which we resembled them, but not with any view of showing that they could exist in fact.

True, he said.

Would a painter be any the worse because, after having delineated with consummate art an ideal of a perfectly beautiful man, he was unable to show that any such man could ever have existed?

He would be none the worse.

Well, and were we not creating an ideal of a perfect State?

To be sure.

And is our theory a worse theory because we are unable to prove the possibility of a city being ordered in the manner described?

Surely not, he replied.

That is the truth, I said. But if, at your request, I am to try and show how and under what conditions the possibility is highest, I must ask you, having this in view, to repeat your former admissions.

What admissions?

[473] I want to know whether ideals are ever fully realized in language? Does not the word express more than the fact, and must not the actual, whatever a man may think, always, in the nature of things, fall short of the truth? What do you say?

I agree.

Then you must not insist on my proving that the actual State will in every respect coincide with the ideal: if we are only able to discover how a city may be governed nearly as we proposed, you will admit that we have discovered the possibility which you demand; and will be contented. I am sure that I should be contented – will not you?

Yes, I will.

Let me next endeavour to show what is that fault in States which is the cause of their present maladministration, and what is the least change which will enable a State to pass into the truer form; and let the change, if possible, be of one thing only, or, if not, of two; at any rate, let the changes be as few and slight as possible.

Certainly, he replied.

I think, I said, that there might be a reform of the State if only one change were made, which is not slight or easy, though still a possible one.

What is it? he said.

Now then, I said, I go to meet that which I liken to the greatest of the waves; yet shall the word be spoken, even though the wave break and drown me in laughter and dishonour; and do you mark my words.

Proceed.

I said: 'Until philosophers are kings, or the kings and princes of this world have the spirit and power of philosophy, and political greatness and wisdom meet in one, and those commoner natures who pursue either to the exclusion of the other are compelled to stand aside, cities will never have rest from their evils, – nor the human race, as I believe, – and then only will this our State have a possibility of life and behold the light of day.' Such was the thought, my dear Glaucon, which I would fain have uttered if it had not seemed too extravagant; for to be convinced that in no other State can there be happiness private or public is indeed a hard thing.

Socrates, what do you mean? I would have you consider that the word which you have uttered is one at which numerous persons, and very [474] respectable persons too, in a figure pulling off their coats all in a moment, and seizing any weapon that comes to hand, will run at you might and main, before you know where you are, intending to do heaven knows what; and if you don't prepare an answer, and put yourself in motion, you will be 'pared by their fine wits,' and no mistake.

You got me into the scrape, I said.

And I was quite right; however, I will do all I can to get you out of it; but I can only give you good-will and good advice, and, perhaps, I may be able to fit answers to your questions better than another – that is all. And now, having such an auxiliary, you must do your best to show the unbelievers that you are right.

I ought to try, I said, since you offer me such invaluable assistance. And I think that, if there is to be a chance of our escaping, we must explain to them whom we mean when we say that philosophers are to rule in the State; then we shall be able to defend ourselves: There will be discovered to be some natures who ought to study philosophy and to be leaders in the State; and others who are not born to be philosophers, and are meant to be followers rather than leaders.

Then now for a definition, he said.

Follow me, I said, and I hope that I may in some way or other be able to give you a satisfactory explanation.

Proceed.

I dare say that you remember, and therefore I need not remind you, that a lover, if he is worthy of the name, ought to show his love, not to someone part of that which he loves, but to the whole.

I really do not understand, and therefore beg of you to assist my memory.

Another person, I said, might fairly reply as you do; but a man of pleasure like yourself ought to know that all who are in the flower of youth do somehow or other raise a pang of emotion in a lover's breast, and are thought by him to be worthy of his affectionate regards. Is not this a way which you have with the fair: one has a snub nose, and you praise his charming face; the hook-nose of another has, you say, a royal look; while he who is neither snub nor hooked has the grace of regularity: the dark visage is manly, the fair are children of the gods; and as to the sweet 'honey pale,' as they are called, what is the very name but the invention of a lover who talks in diminutives, and is not averse to paleness if appearing on the cheek of youth? In a word, [475] there is no excuse which you will not make, and nothing which you will not say, in order not to lose a single flower that blooms in the spring-time of youth.

If you make me an authority in matters of love, for the sake of the argument, I assent.

And what do you say of lovers of wine? Do you not see them doing the same? They are glad of any pretext of drinking any wine.

Very good.

And the same is true of ambitious men; if they cannot command an army, they are willing to command a file; and if they cannot be honoured by really great and important persons, they are glad to be honoured by lesser and meaner people, – but honour of some kind they must have.

Exactly.

Once more let me ask: Does he who desires any class of goods, desire the whole class or a part only?

The whole.

And may we not say of the philosopher that he is a lover, not of a part of wisdom only, but of the whole?

Yes, of the whole.

And he who dislikes learning, especially in youth, when he has no power of judging what is good and what is not, such an one we maintain not to be a philosopher or a lover of knowledge, just as he who refuses his food is not hungry, and may be said to have a bad appetite and not a good one?

Very true, he said.

Whereas he who has a taste for every sort of knowledge and who is curious to learn and is never satisfied, may be justly termed a philosopher? Am I not right?

Glaucon said: If curiosity makes a philosopher, you will find many a strange being will have a title to the name. All the lovers of sights have a delight in learning, and must therefore be included. Musical amateurs, too, are a folk strangely out of place among philosophers, for they are the last persons in the world who would come to anything like a philosophical discussion, if they could help, while they run about at the Dionysiac festivals as if they had let out their ears to hear every chorus; whether the performance is in town or country – that makes no difference – they are there. Now are we to maintain that all these and any who have similar tastes, as well as the professors of quite minor arts, are philosophers?

Certainly not, I replied; they are only an imitation.

He said: Who then are the true philosophers?

Those, I said, who are lovers of the vision of truth.

That is also good, he said; but I should like to know what you mean?

To another, I replied, I might have a difficulty in explaining; but I am sure that you will admit a proposition which I am about to make.

What is the proposition?

[476] That since beauty is the opposite of ugliness, they are two?

Certainly.

And inasmuch as they are two, each of them is one?

True again.

And of just and unjust, good and evil, and of every other class, the same remark holds: taken singly, each of them is one; but from the various combinations of them with actions and things and with one another, they are seen in all sorts of lights and appear many?

Very true.

And this is the distinction which I draw between the sight-loving, art-loving, practical class and those of whom I am speaking, and who are alone worthy of the name of philosophers.

How do you distinguish them? he said.

The lovers of sounds and sights, I replied, are, as I conceive, fond of fine tones and colours and forms and all the artificial products that are made out of them, but their mind is incapable of seeing or loving absolute beauty.

True, he replied.

Few are they who are able to attain to the sight of this.

Very true.

And he who, having a sense of beautiful things has no sense of absolute beauty, or who, if another lead him to a knowledge of that beauty is unable to follow – of such an one I ask, Is he awake or in a dream only? Reflect: is not the dreamer, sleeping or waking, one who likens dissimilar things, who puts the copy in the place of the real object?

I should certainly say that such an one was dreaming.

But take the case of the other, who recognises the existence of absolute beauty and is able to distinguish the idea from the objects which participate in the idea, neither putting the objects in the place of the idea nor the idea in the place of the objects – is he a dreamer, or is he awake?

He is wide awake.

And may we not say that the mind of the one who knows has knowledge, and that the mind of the other, who opines only, has opinion?

Certainly.

But suppose that the latter should quarrel with us and dispute our statement, can we administer any soothing cordial or advice to him, without revealing to him that there is sad disorder in his wits?

We must certainly offer him some good advice, he replied.

Come, then, and let us think of something to say to him. Shall we begin by assuring him that he is welcome to any knowledge which he may have, and that we are rejoiced at his having it? But we should like to ask him a question: Does he who has knowledge know something or nothing? (You must answer for him.)

I answer that he knows something.

Something that is or is not?

[477] Something that is; for how can that which is not ever be known?

And are we assured, after looking at the matter from many points of view, that absolute being is or may be absolutely known, but that the utterly non-existent is utterly unknown?

Nothing can be more certain.

Good. But if there be anything which is of such a nature as to be and not to be, that will have a place intermediate between pure being and the absolute negation of being?

Yes, between them.

And, as knowledge corresponded to being and ignorance of necessity to not-being, for that intermediate between being and not-being there has to be discovered a corresponding intermediate between ignorance and knowledge, if there be such?

Certainly.

Do we admit the existence of opinion?

Undoubtedly.

As being the same with knowledge, or another faculty?

Another faculty.

Then opinion and knowledge have to do with different kinds of matter corresponding to this difference of faculties?

Yes.

And knowledge is relative to being and knows being. But before I proceed further I will make a division.

What division?

I will begin by placing faculties in a class by themselves: they are powers in us, and in all other things, by which we do as we do. Sight and hearing, for example, I should call faculties. Have I clearly explained the class which I mean?

Yes, I quite understand.

Then let me tell you my view about them. I do not see them, and therefore the distinctions of figure, colour, and the like, which enable me to discern the differences of some things, do not apply to them. In speaking of a faculty I think only of its sphere and its result; and that which has the same sphere and the same result I call the same faculty, but that which has another sphere and another result I call different. Would that be your way of speaking?

Yes.

And will you be so very good as to answer one more question? Would you say that knowledge is a faculty, or in what class would you place it?

Certainly knowledge is a faculty, and the mightiest of all faculties.

And is opinion also a faculty?

Certainly, he said; for opinion is that with which we are able to form an opinion.

And yet you were acknowledging a little while ago that knowledge is not the same as opinion?

Why, yes, he said: how can any reasonable being ever identify that which is infallible with that which errs?

[478] An excellent answer, proving, I said, that we are quite conscious of a distinction between them.

Yes.

Then knowledge and opinion, having distinct powers, have also distinct spheres or subject-matters?

That is certain.

Being is the sphere or subject-matter of knowledge, and knowledge is to know the nature of being?

Yes.

And opinion is to have an opinion?

Yes.

And do we know what we opine? Or is the subject-matter of opinion the same as the subject-matter of knowledge?

Nay, he replied, that has been already disproven; if difference in faculty implies difference in the sphere or subject-matter, and if, as we were saying, opinion and knowledge are distinct faculties, then the sphere of knowledge and of opinion cannot be the same.

Then if being is the subject-matter of knowledge, something else must be the subject-matter of opinion?

Yes, something else.

Well then, is not-being the subject-matter of opinion? Or, rather, how can there be an opinion at all about not-being? Reflect: when a man has an opinion, has he not an opinion about something? Can he have an opinion which is an opinion about nothing?

Impossible.

He who has an opinion has an opinion about some one thing?

Yes.

And not-being is not one thing but, properly speaking, nothing?

True.

Of not-being, ignorance was assumed to be the necessary correlative; of being, knowledge?

True, he said.

Then opinion is not concerned either with being or with not-being?

Not with either.

And can therefore neither be ignorance nor knowledge?

That seems to be true.

But is opinion to be sought without and beyond either of them, in a greater clearness than knowledge, or in a greater darkness than ignorance?

In neither.

Then I suppose that opinion appears to you to be darker than knowledge, but lighter than ignorance?

Both; and in no small degree.

And also to be within and between them?

Yes.

Then you would infer that opinion is intermediate?

No question.

But were we not saying before, that if anything appeared to be of a sort which is and is not at the same time, that sort of thing would appear also to lie in the interval between pure being and absolute not-being; and that the corresponding faculty is neither knowledge nor ignorance, but will be found in the interval between them?

True.

And in that interval there has now been discovered something which we call opinion?

There has.

Then what remains to be discovered is the object which partakes equally of the nature of being and not-being, and cannot rightly be termed either, pure and simple; this unknown term, when discovered, we may truly call the subject of opinion, and assign each to their proper faculty, – the extremes to the faculties of the extremes and the mean to the faculty of the mean.

True.

This being premised, I would ask [479] the gentleman who is of opinion that there is no absolute or unchangeable idea of beauty – in whose opinion the beautiful is the manifold – he, I say, your lover of beautiful sights, who cannot bear to be told that the beautiful is one, and the just is one, or that anything is one – to him I would appeal, saying, Will you be so very kind, sir, as to tell us whether, of all these beautiful things, there is one which will not be found ugly; or of the just, which will not be found unjust; or of the holy, which will not also be unholy?

No, he replied; the beautiful will in some point of view be found ugly; and the same is true of the rest.

And may not the many which are doubles be also halves? – doubles, that is, of one thing, and halves of another?

Quite true.

And things great and small, heavy and light, as they are termed, will not be denoted by these any more than by the opposite names?

True; both these and the opposite names will always attach to all of them.

And can any one of those many things which are called by particular names be said to be this rather than not to be this?

He replied: They are like the punning riddles which are asked at feasts or the children's puzzle about the eunuch aiming at the bat, with what he hit him, as they say in the puzzle, and upon what the bat was sitting. The individual objects of which I am speaking are also a riddle, and have a double sense: nor can you fix them in your mind, either as being or not-being, or both, or neither.

Then what will you do with them? I said. Can they have a better place than between being and not-being? For they are clearly not in greater darkness or negation than not-being, or more full of light and existence than being.

That is quite true, he said.

Thus then we seem to have discovered that the many ideas which the multitude entertain about the beautiful and about all other things are tossing about in some region which is half-way between pure being and pure not-being?

We have.

Yes; and we had before agreed that anything of this kind which we might find was to be described as matter of opinion, and not as matter of knowledge; being the intermediate flux which is caught and detained by the intermediate faculty.

Quite true.

Then those who see the many beautiful, and who yet neither see absolute beauty, nor can follow any guide who points the way thither; who see the many just, and not absolute justice, and the like, – such persons may be said to have opinion but not knowledge?

That is certain.

But those who see the absolute and eternal and immutable may be said to know, and not to have opinion only?

Neither can that be denied.

The one love and embrace the subjects of [480] knowledge, the other those of opinion? The latter are the same, as I dare say you will remember, who listened to sweet sounds and gazed upon fair colours, but would not tolerate the existence of absolute beauty.

Yes, I remember.

Shall we then be guilty of any impropriety in calling them lovers of opinion rather than lovers of wisdom, and will they be very angry with us for thus describing them?

I shall tell them not to be angry; no man should be angry at what is true.

But those who love the truth in each thing are to be called lovers of wisdom and not lovers of opinion.

Assuredly.

BOOK VI

[484a-485a omitted]

[485]

Let us suppose that philosophical minds always love knowledge of a sort which shows them the eternal nature not varying from generation and corruption.

Agreed.

And further, I said, let us agree that they are lovers of all true being; there is no part whether greater or less, or more or less honourable, which they are willing to renounce; as we said before of the lover and the man of ambition.

True.

And if they are to be what we were describing, is there not another quality which they should also possess?

What quality?

Truthfulness: they will never intentionally receive into their mind falsehood, which is their detestation, and they will love the truth.

Yes, that may be safely affirmed of them.

'Maybe,' my friend, I replied, is not the word; say rather 'must be affirmed' for he whose nature is amorous of anything cannot help loving all that belongs or is akin to the object of his affections.

Right, he said.

And is there anything more akin to wisdom than truth?

How can there be?

Can the same nature be a lover of wisdom and a lover of falsehood?

Never.

The true lover of learning then must from his earliest youth, as far as in him lies, desire all truth?

Assuredly.

But then again, as we know by experience, he whose desires are strong in one direction will have them weaker in others; they will be like a stream which has been drawn off into another channel.

True.

He whose desires are drawn towards knowledge in every form will be absorbed in the pleasures of the soul, and will hardly feel bodily pleasure – I mean, if he be a true philosopher and not a sham one.

That is most certain.

Such a one is sure to be temperate and the reverse of covetous; for the motives which make another man desirous of having and spending, have no place in his character.

Very true.

[486] Another criterion of the philosophical nature has also to be considered.

What is that?

There should be no secret corner of illiberality; nothing can be more antagonistic than meanness to a soul which is ever longing after the whole of things both divine and human.

Most true, he replied.

Then how can he who has magnificence of mind and is the spectator of all time and all existence, think much of human life?

He cannot.

Or can such a one account death fearful?

No indeed.

Then the cowardly and mean nature has no part in true philosophy?

Certainly not.

Or again: can he who is harmoniously constituted, who is not covetous or mean, or a boaster, or a coward – can he, I say, ever be unjust or hard in his dealings?

Impossible.

Then you will soon observe whether a man is just and gentle, or rude and unsociable; these are the signs which distinguish even in youth the philosophical nature from the unphilosophical.

True.

There is another point which should be remarked.

What point?

Whether he has or has not a pleasure in learning; for no one will love that which gives him pain, and in which after much toil he makes little progress.

Certainly not.

And again, if he is forgetful and retains nothing of what he learns, will he not be an empty vessel?

That is certain.

Labouring in vain, he must end in hating himself and his fruitless occupation?

Yes.

Then a soul which forgets cannot be ranked among genuine philosophic natures; we must insist that the philosopher should have a good memory?

Certainly.

And once more, the inharmonious and unseemly nature can only tend to disproportion?

Undoubtedly.

And do you consider truth to be akin to proportion or to disproportion?

To proportion.

Then, besides other qualities, we must try to find a naturally well-proportioned and gracious mind, which will move spontaneously towards the true being of everything.

Certainly.

Well, and do not all these qualities, which we have been enumerating, go together, and are they not, in a manner, necessary to a soul, which is to have a full and perfect participation of being?

[487] They are absolutely necessary, he replied.

And must not that be a blameless study which he only can pursue who has the gift of a good memory, and is quick to learn, – noble, gracious, the friend of truth, justice, courage, temperance, who are his kindred?

The god of jealousy himself, he said, could find no fault with such a study.

And to men like him, I said, when perfected by years and education, and to these only you will entrust the State.

[487b-504c has been omitted. Plato here seeks to explain why the public's perception of the philosopher is so at odds with the responsibilities afforded to him or her in the state that has been introduced. The circumstances in which any actual philosopher has been raised are quite insufficient. Even so he thinks that the correct education of the philosopher in the correct state would confirm that his theoretical state is possible.]

The guardian then, I said, must be required to take the longer circuit, and toil at learning as well as at gymnastics, or he will never reach the highest knowledge of all which, as we were just now saying, is his proper calling.

What, he said, is there a knowledge still higher than this – higher than justice and the other virtues?

Yes, I said, there is. And of the virtues too we must behold not the outline merely, as at present – nothing short of the most finished picture should satisfy us. When little things are elaborated with an infinity of pains, in order that they may appear in their full beauty and utmost clearness, how ridiculous that we should not think the highest truths worthy of attaining the highest accuracy!

A right noble thought; but do you suppose that we shall refrain from asking you what is this highest knowledge?

Nay, I said, ask if you will; but I am certain that you have heard the answer many times, and now you either do not understand me or, as I rather think, you are disposed [505] to be troublesome; for you have often been told that the idea of good is the highest knowledge, and that all other things become useful and advantageous only by their use of this. You can hardly be ignorant that of this I was about to speak, concerning which, as you have often heard me say, we know so little; and, without which, any other knowledge or possession of any kind will profit us nothing. Do you think that the possession of all other things is of any value if we do not possess the good? Or the knowledge of all other things if we have no knowledge of beauty and goodness?

Assuredly not.

You are further aware that most people affirm pleasure to be the good, but the finer sort of wits say it is knowledge?

Yes.

And you are aware too that the latter cannot explain what they mean by knowledge, but are obliged after all to say knowledge of the good?

How ridiculous!

Yes, I said, that they should begin by reproaching us with our ignorance of the good, and then presume our knowledge of it – for the good they define to be knowledge of the good, just as if we understood them when they use the term 'good' – this is of course ridiculous.

Most true, he said.

And those who make pleasure their good are in equal perplexity; for they are compelled to admit that there are bad pleasures as well as good.

Certainly.

And therefore to acknowledge that bad and good are the same?

True.

There can be no doubt about the numerous difficulties in which this question is involved.

There can be none.

Further, do we not see that many are willing to do or to have or to seem to be what is just and honourable without the reality; but no one is satisfied with the appearance of good – the reality is what they seek; in the case of the good, appearance is despised by everyone.

Very true, he said.

Of this then, which every soul of man pursues and makes the end of all his actions, having a presentiment that there is such an end, and yet hesitating because neither knowing the nature nor having the same assurance of this as of other things, and therefore losing whatever good there is in other things, – [506] of a principle such and so great as this ought the best men in our State, to whom everything is entrusted, to be in the darkness of ignorance?

Certainly not, he said.

I am sure, I said, that he who does not know how the beautiful and the just are likewise good will be but a sorry guardian of them; and I suspect that no one who is ignorant of the good will have a true knowledge of them.

That, he said, is a shrewd suspicion of yours.

And if we only have a guardian who has this knowledge our State will be perfectly ordered?

Of course, he replied; but I wish that you would tell me whether you conceive this supreme principle of the good to be knowledge or pleasure, or different from either?

Aye, I said, I knew all along that a fastidious gentleman like you would not be contented with the thoughts of other people about these matters.

True, Socrates; but I must say that one who like you has passed a lifetime in the study of philosophy should not be always repeating the opinions of others, and never telling his own.

Well, but has anyone a right to say positively what he does not know?

Not, he said, with the assurance of positive certainty; he has no right to do that: but he may say what he thinks, as a matter of opinion.

And do you not know, I said, that all mere opinions are bad, and the best of them blind? You would not deny that those who have any true notion without intelligence are only like blind men who feel their way along the road?

Very true.

And do you wish to behold what is blind and crooked and base, when others will tell you of brightness and beauty?

Still, I must implore you, Socrates, said Glaucon, not to turn away just as you are reaching the goal; if you will only give such an explanation of the good as you have already given of justice and temperance and the other virtues, we shall be satisfied.

Yes, my friend, and I shall be at least equally satisfied, but I cannot help fearing that I shall fail, and that my indiscreet zeal will bring ridicule upon me. No, sweet sirs, let us not at present ask what is the actual nature of the good, for to reach what is now in my thoughts would be an effort too great for me. But of the child of the good who is likest him, I would fain speak, if I could be sure that you wished to hear – otherwise, not.

By all means, he said, tell us about the child, and you shall remain in our debt for the account of the parent.

[507] I do indeed wish, I replied, that I could pay, and you receive, the account of the parent, and not, as now, of the offspring only; take, however, this latter by way of interest, and at the same time have a care that I do not render a false account, although I have no intention of deceiving you.

Yes, we will take all the care that we can: proceed.

Yes, I said, but I must first come to an understanding with you, and remind you of what I have mentioned in the course of this discussion, and at many other times.

What?

The old story, that there is a many beautiful and a many good, and so of other things which we describe and define; to all of them the term 'many' is applied.

True, he said.

And there is an absolute beauty and an absolute good, and of other things to which the term 'many' is applied there is an absolute; for they may be brought under a single idea, which is called the essence of each.

Very true.

The many, as we say, are seen but not known, and the ideas are known but not seen.

Exactly.

And what is the organ with which we see the visible things?

The sight, he said.

And with the hearing, I said, we hear, and with the other senses perceive the other objects of sense?

True.

But have you remarked that sight is by far the most costly and complex piece of workmanship which the artificer of the senses ever contrived?

No, I never have, he said.

Then reflect; has the ear or voice need of any third or additional nature in order that the one may be able to hear and the other to be heard?

Nothing of the sort.

No, indeed, I replied; and the same is true of most, if not all, the other senses – you would not say that any of them requires such an addition?

Certainly not.

But you see that without the addition of some other nature there is no seeing or being seen?

How do you mean?

Sight being, as I conceive, in the eyes, and he who has eyes wanting to see; colour being also present in them, still unless there be a third nature specially adapted to the purpose, the owner of the eyes will see nothing and the colours will be invisible.

Of what nature are you speaking?

Of that which you term light, I replied.

True, he said.

Noble, then, is the bond which links together sight and visibility, [508] and great beyond other bonds by no small difference of nature; for light is their bond, and light is no ignoble thing?

Nay, he said, the reverse of ignoble.

And which, I said, of the gods in heaven would you say was the lord of this element? Whose is that light which makes the eye to see perfectly and the visible to appear?

You mean the sun, as you and all mankind say.

May not the relation of sight to this deity be described as follows?

How?

Neither sight nor the eye in which sight resides is the sun?

No.

Yet of all the organs of sense the eye is the most like the sun?

By far the most like.

And the power which the eye possesses is a sort of effluence which is dispensed from the sun?

Exactly.

Then the sun is not sight, but the author of sight who is recognised by sight?

True, he said.

And this is he whom I call the child of the good, whom the good begat in his own likeness, to be in the visible world, in relation to sight and the things of sight, what the good is in the intellectual world in relation to mind and the things of mind.

Will you be a little more explicit? he said.

Why, you know, I said, that the eyes, when a person directs them towards objects on which the light of day is no longer shining, but the moon and stars only, see dimly, and are nearly blind; they seem to have no clearness of vision in them?

Very true.

But when they are directed towards objects on which the sun shines, they see clearly and there is sight in them?

Certainly.

And the soul is like the eye: when resting upon that on which truth and being shine, the soul perceives and understands, and is radiant with intelligence; but when turned towards the twilight of becoming and perishing, then she has opinion only, and goes blinking about, and is first of one opinion and then of another, and seems to have no intelligence?

Just so.

Now, that which imparts truth to the known and the power of knowing to the knower is what I would have you term the idea of good, and this you will deem to be the cause of science, and of truth insofar as the latter becomes the subject of knowledge; beautiful too, as are both truth and knowledge, you will be right in esteeming this other nature as more beautiful than either; [509] and, as in the previous instance, light and sight may be truly said to be like the sun, and yet not to be the sun, so in this other sphere, science and truth may be deemed to be like the good, but not the good; the good has a place of honour yet higher.

What a wonder of beauty that must be, he said, which is the author of science and truth, and yet surpasses them in beauty; for you surely cannot mean to say that pleasure is the good?

God forbid, I replied; but may I ask you to consider the image in another point of view?

In what point of view?

You would say, would you not, that the sun is not only the author of visibility in all visible things, but of generation and nourishment and growth, though he himself is not generation?

Certainly.

In like manner the good may be said to be not only the author of knowledge to all things known, but of their being and essence, and yet the good is not essence, but far exceeds essence in dignity and power.

Glaucon said, with a ludicrous earnestness: By the light of heaven, how amazing!

Yes, I said, and the exaggeration may be set down to you; for you made me utter my fancies.

[509c-511e introduces the more technical and puzzling Divided Line. This is here omitted, but one can get a sense of it from 533-534 in the following Book.]

BOOK VII

[514]

And now, I said, let me show in a figure how far our nature is enlightened or unenlightened: – Behold! human beings living in a underground den, which has a mouth open towards the light and reaching all along the den; here they have been from their childhood, and have their legs and necks chained so that they cannot move, and can only see before them, being prevented by the chains from turning round their heads. Above and behind them a fire is blazing at a distance, and between the fire and the prisoners there is a raised way; and you will see, if you look, a low wall built along the way, like the screen which marionette players have in front of them, over which they show the puppets.

I see.

And do you see, I said, men passing along the wall carrying all sorts of vessels, and statues and [515] figures of animals made of wood and stone and various materials, which appear over the wall? Some of them are talking, others silent.

You have shown me a strange image, and they are strange prisoners.

Like ourselves, I replied; and they see only their own shadows, or the shadows of one another, which the fire throws on the opposite wall of the cave?

True, he said; how could they see anything but the shadows if they were never allowed to move their heads?

And of the objects which are being carried in like manner they would only see the shadows?

Yes, he said.

And if they were able to converse with one another, would they not suppose that they were naming what was actually before them?

Very true.

And suppose further that the prison had an echo which came from the other side, would they not be sure to fancy when one of the passers-by spoke that the voice which they heard came from the passing shadow?

No question, he replied.

To them, I said, the truth would be literally nothing but the shadows of the images.

That is certain.

And now look again, and see what will naturally follow if the prisoners are released and disabused of their error. At first, when any of them is liberated and compelled suddenly to stand up and turn his neck round and walk and look towards the light, he will suffer sharp pains; the glare will distress him, and he will be unable to see the realities of which in his former state he had seen the shadows; and then conceive someone saying to him, that what he saw before was an illusion, but that now, when he is approaching nearer to being and his eye is turned towards more real existence, he has a clearer vision, – what will be his reply? And you may further imagine that his instructor is pointing to the objects as they pass and requiring him to name them, – will he not be perplexed? Will he not fancy that the shadows which he formerly saw are truer than the objects which are now shown to him?

Far truer.

And if he is compelled to look straight at the light, will he not have a pain in his eyes which will make him turn away to take refuge in the objects of vision which he can see, and which he will conceive to be in reality clearer than the things which are now being shown to him?

True, he said.

And suppose once more, that he is reluctantly dragged up a steep and rugged ascent, and held fast until he is forced into the presence of the sun himself, is he not likely to be pained and [516] irritated? When he approaches the light his eyes will be dazzled, and he will not be able to see anything at all of what are now called realities.

Not all in a moment, he said.

He will require to grow accustomed to the sight of the upper world. And first he will see the shadows best, next the reflections of men and other objects in the water, and then the objects themselves; then he will gaze upon the light of the moon and the stars and the spangled heaven; and he will see the sky and the stars by night better than the sun or the light of the sun by day?

Certainly.

Last of all he will be able to see the sun, and not mere reflections of him in the water, but he will see him in his own proper place, and not in another; and he will contemplate him as he is.

Certainly.

He will then proceed to argue that this is he who gives the season and the years, and is the guardian of all that is in the visible world, and in a certain way the cause of all things which he and his fellows have been accustomed to behold?

Clearly, he said, he would first see the sun and then reason about him.

And when he remembered his old habitation, and the wisdom of the den and his fellow-prisoners, do you not suppose that he would felicitate himself on the change, and pity them?

Certainly, he would.

And if they were in the habit of conferring honours among themselves on those who were quickest to observe the passing shadows and to remark which of them went before, and which followed after, and which were together; and who were therefore best able to draw conclusions as to the future, do you think that he would care for such honours and glories, or envy the possessors of them? Would he not say with Homer,

'Better to be the poor servant of a poor master,'

And to endure anything, rather than think as they do and live after their manner?

Yes, he said, I think that he would rather suffer anything than entertain these false notions and live in this miserable manner.

Imagine once more, I said, such a one coming suddenly out of the sun to be replaced in his old situation; would he not be certain to have his eyes full of darkness?

To be sure, he said.

And if there were a contest, and he had to compete in measuring the shadows with the prisoners who had never moved out of the den, while his sight was still weak, [517] and before his eyes had become steady (and the time which would be needed to acquire this new habit of sight might be very considerable), would he not be ridiculous? Men would say of him that up he went and down he came without his eyes; and that it was better not even to think of ascending; and if anyone tried to loose another and lead him up to the light, let them only catch the offender, and they would put him to death.

No question, he said.

This entire allegory, I said, you may now append, dear Glaucon, to the previous argument; the prison-house is the world of sight, the light of the fire is the sun, and you will not misapprehend me if you interpret the journey upwards to be the ascent of the soul into the intellectual world according to my poor belief, which, at your desire, I have expressed – whether rightly or wrongly God knows. But, whether true or false, my opinion is that in the world of knowledge the idea of good appears last of all, and is seen only with an effort; and, when seen, is also inferred to be the universal author of all things beautiful and right, parent of light and of the lord of light in this visible world, and the immediate source of reason and truth in the intellectual; and that this is the power upon which he who would act rationally either in public or private life must have his eye fixed.

I agree, he said, as far as I am able to understand you.

Moreover, I said, you must not wonder that those who attain to this beatific vision are unwilling to descend to human affairs; for their souls are ever hastening into the upper world where they desire to dwell; which desire of theirs is very natural, if our allegory may be trusted.

Yes, very natural.

And is there anything surprising in one who passes from divine contemplations to the evil state of man, misbehaving himself in a ridiculous manner; if, while his eyes are blinking and before he has become accustomed to the surrounding darkness, he is compelled to fight in courts of law, or in other places, about the images or the shadows of images of justice, and is endeavouring to meet the conceptions of those who have never yet seen absolute justice?

Anything but surprising, he replied.

[518] Anyone who has common sense will remember that the bewilderments of the eyes are of two kinds, and arise from two causes, either from coming out of the light or from going into the light, which is true of the mind's eye, quite as much as of the bodily eye; and he who remembers this when he sees anyone whose vision is perplexed and weak, will not be too ready to laugh; he will first ask whether that soul of man has come out of the brighter life, and is unable to see because

unaccustomed to the dark, or having turned from darkness to the day is dazzled by excess of light. And he will count the one happy in his condition and state of being, and he will pity the other; or, if he have a mind to laugh at the soul which comes from below into the light, there will be more reason in this than in the laugh which greets him who returns from above out of the light into the den.

That, he said, is a very just distinction.

But then, if I am right, certain professors of education must be wrong when they say that they can put a knowledge into the soul which was not there before, like sight into blind eyes.

They undoubtedly say this, he replied.

Whereas, our argument shows that the power and capacity of learning exists in the soul already; and that just as the eye was unable to turn from darkness to light without the whole body, so too the instrument of knowledge can only by the movement of the whole soul be turned from the world of becoming into that of being, and learn by degrees to endure the sight of being, and of the brightest and best of being, or in other words, of the good.

Very true.

And must there not be some art which will effect conversion in the easiest and quickest manner; not implanting the faculty of sight, for that exists already, but has been turned in the wrong direction, and is looking away from the truth?

[In 518d-531e, omitted here, the five mathematical sciences, arithmetic, geometry, stereometry, music

and astronomy, if practised theoretically, are identified with this art. But another art is needed to approach the Good.]

[532a] And so, Glaucon, I said, we have at last arrived at the hymn of dialectic. This is that strain which is of the intellect only, but which the faculty of sight will nevertheless be found to imitate; for sight, as you may remember, was imagined by us after a while to behold the real animals and stars, and last of all the sun himself. And so with dialectic; when a person starts on the discovery of the absolute by the light of reason only, and without any assistance of sense, and perseveres until by pure intelligence he arrives at the perception of the absolute good, he at last finds himself at the end of the intellectual world, as in the case of sight at the end of the visible.

Exactly, he said.

Then this is the progress which you call dialectic?

True.

But the release of the prisoners from chains, and their translation from the shadows to the images and to the light, and the ascent from the underground den to the sun, while in his presence they are vainly trying to look on animals and plants and the light of the sun, but are able to perceive even with their weak eyes the images in the water (which are divine), and are the shadows of true existence (not shadows of images cast by a light of fire, which compared with the sun is only an image) – this power of elevating the highest principle in the soul to the contemplation of that which is best in existence, with which we may compare the raising of that faculty which is the very light of the body to the sight of that which is brightest in the material and visible world – this power is given, as I was saying, by all that study and pursuit of the arts which has been described.

I agree in what you are saying, he replied.

[minor omissions, 532d-533c, 533d-e]

Then dialectic, and dialectic alone, goes directly to the first principle and is the only science which does away with hypotheses in order to make her ground secure; the eye of the soul, which is literally buried in an outlandish slough, is by her gentle aid lifted upwards; and she uses as handmaids and helpers in the work of conversion, the sciences which we have been discussing. .... At any rate, we are satisfied, as before (at 509e-511e), to have four divisions; two for intellect and two for opinion, and to call the first division science, the second understanding, [534] the third belief, and the fourth perception of shadows, opinion being concerned with becoming, and intellect with being; and so to make a proportion: –

As being is to becoming, so is pure intellect to opinion. And as intellect is to opinion, so is science to belief, and understanding to the perception of shadows.

But let us defer the further correlation and subdivision of the subjects of opinion and of intellect, for it will be a long enquiry, many times longer than this has been.

As far as I understand, he said, I agree.

And do you also agree, I said, in describing the dialectician as one who attains a conception of the essence of each thing?

And he who does not possess and is therefore unable to impart this conception, in whatever degree he fails, may in that degree also be said to fail in intelligence? Will you admit so much?

Yes, he said; how can I deny it?

And you would say the same of the conception of the good? Until the person is able to abstract and define rationally the idea of good, and unless he can run the gauntlet of all objections, and is ready to disprove them, not by appeals to opinion, but to absolute truth, never faltering at any step of the argument – unless he can do all this, you would say that he knows neither the idea of good nor any other good; he apprehends only a shadow, if anything at all, which is given by opinion and not by science; – dreaming and slumbering in this life, before he is well awake here, he arrives at the world below, and has his final quietus.

In all that I should most certainly agree with you.

[minor omission, 534d-e]

Dialectic, then, as you will agree, is the coping-stone of the sciences, and is set over them; no other science can be placed higher – [535] the nature of knowledge can no further go?

I agree, he said.

But to whom we are to assign these studies, and in what way they are to be assigned, are questions which remain to be considered.

Yes, clearly.

You remember, I said, how the rulers were chosen before?

Certainly, he said.

The same natures must still be chosen, and the preference again given to the surest and the bravest, and, if possible, to the fairest; and, having noble and generous tempers, they should also have the natural gifts which will facilitate their education.

[535b-536d omitted]

And, therefore, calculation and geometry and all the other elements of instruction, which are a preparation for dialectic, should be presented to the mind in childhood; not, however, under any notion of forcing our system of education.

Why not?

Because a freeman ought not to be a slave in the acquisition of knowledge of any kind. Bodily exercise, when compulsory, does no harm to the body; but knowledge which is acquired under compulsion obtains no hold on the mind.

Very true.

[minor omission, 536e-537c]

These, I said, are the points which you must consider; and those who have most of this comprehension, and who are most steadfast in their learning, and in their military and other appointed duties, when they have arrived at the age of thirty have to be chosen by you out of the select class, and elevated to higher honour; and you will have to prove them by the help of dialectic, in order to learn which of them is able to give up the use of sight and the other senses, and in company with truth to attain absolute being: And here, my friend, great caution is required.

Why great caution?

Do you not remark, I said, how great is the evil which dialectic has introduced?

What evil? he said.

The students of the art are filled with lawlessness.

Quite true, he said.

[537e-539a omitted]

Therefore, that your feelings may not be moved to pity about our citizens who are now thirty years of age, every care must be taken in introducing them to dialectic.

Certainly.

There is a danger lest they should taste the dear delight too early; for youngsters, as you may have observed, when they first get the taste in their mouths, argue for amusement, and are always contradicting and refuting others in imitation of those who refute them; like puppy-dogs, they rejoice in pulling and tearing at all who come near them.

Yes, he said, there is nothing which they like better.

[minor omission, 539b-d]

And did we not make special provision for this, when we said that the disciples of philosophy were to be orderly and steadfast, not, as now, any chance aspirant or intruder?

Very true.

Suppose, I said, the study of philosophy to take the place of gymnastics and to be continued diligently and earnestly and exclusively for twice the number of years which were passed in bodily exercise – will that be enough?

Would you say six or four years? he asked.

Say five years, I replied; at the end of the time they must be sent down again into the den and compelled to hold any military or other office which young men are qualified to hold:

in this way they will get their experience of life, and there will be an opportunity of trying whether, [540] when they are drawn all manner of ways by temptation, they will stand firm or flinch.

And how long is this stage of their lives to last?

Fifteen years, I answered; and when they have reached fifty years of age, then let those who still survive and have distinguished themselves in every action of their lives and in every branch of knowledge come at last to their consummation: the time has now arrived at which they must raise the eye of the soul to the universal light which lightens all things, and behold the absolute good; for that is the pattern according to which they are to order the State and the lives of individuals, and the remainder of their own lives also; making philosophy their chief pursuit, but, when their turn comes, toiling also at politics and ruling for the public good, not as though they were performing some heroic action, but simply as a matter of duty; and when they have brought up in each generation others like themselves and left them in their place to be governors of the State, then they will depart to the Islands of the Blest and dwell there; and the city will give them public memorials and sacrifices and honour them, if the Pythian oracle consent, as demigods, but if not, as in any case blessed and divine.

You are a sculptor, Socrates, and have made statues of our governors faultless in beauty.

Yes, I said, Glaucon, and of our governesses too; for you must not suppose that what I have been saying applies to men only and not to women as far as their natures can go.

There you are right, he said, since we have made them to share in all things like the men.

Well, I said, and you would agree (would you not?) that what has been said about the State and the government is not a mere dream, and although difficult not impossible, but only possible in the way which has been supposed; that is to say, when the true philosopher kings are born in a State, one or more of them, despising the honours of this present world which they deem mean and worthless, esteeming above all things right and the honour that springs from right, and regarding justice as the greatest and most necessary of all things, whose ministers they are, and whose principles will be exalted by them when they set in order their own city?

[minor omission, 540e-541b]

BOOK VIII

[The principal subject of Book VIII is the decline of the state described into an honour-loving constitution (timocracy), then a wealth-loving one (oligarchy), then into an excessively freedom-loving one (democracy), and so to tyranny. The demolition of the best state seems somehow to mirror its construction in earlier books.]

BOOK IX

[Most of Book IX discusses the nature of the tyrannical man, and this seems to counterbalance the discussion of the philosopher-king in Book VI. This leads naturally on to a pseudo-mathematical calculation purporting to show that the king is 729 times more fortunate than the tyrant! A new section now looks unrelated to what preceded in Books VIII-IX, and seems already like a plausible conclusion to the work as a whole.]

[588]

Well, I said, and now having arrived at this stage of the argument, we may revert to the words which brought us hither: Was not someone saying that injustice was a gain to the perfectly unjust who was reputed to be just?

Yes, that was said.

Now then, having determined the power and quality of justice and injustice, let us have a little conversation with him.

What shall we say to him?

Let us make an image of the soul, that he may have his own words presented before his eyes.

Of what sort?

An ideal image of the soul, like the composite creations of ancient mythology, such as the Chimera or Scylla or Cerberus, and there are many others in which two or more different natures are said to grow into one.

There are said to have been such unions.

Then do you now model the form of a multitudinous, many-headed monster, having a ring of heads of all manner of beasts, tame and wild, which he is able to generate and metamorphose at will.

You suppose marvellous powers in the artist; but, as language is more pliable than wax or any similar substance, let there be such a model as you propose.

Suppose now that you make a second form as of a lion, and a third of a man, the second smaller than the first, and the third smaller than the second.

That, he said, is an easier task; and I have made them as you say.

And now join them, and let the three grow into one.

That has been accomplished.

Next fashion the outside of them into a single image, as of a man, so that he who is not able to look within, and sees only the outer hull, may believe the beast to be a single human creature.

I have done so, he said.

And now, to him who maintains that it is profitable for the human creature to be unjust, and unprofitable to be just, let us reply that, if he be right, it is profitable for this creature to feast the multitudinous monster and strengthen the lion and the lion-like qualities, but to [589] starve and weaken the man, who is consequently liable to be dragged about at the mercy of either of the other two; and he is not to attempt to familiarize or harmonize them with one another – he ought rather to suffer them to fight and bite and devour one another.

Certainly, he said; that is what the approver of injustice says.

To him the supporter of justice makes answer that he should ever so speak and act as to give the man within him in some way or other the most complete mastery over the entire human creature. He should watch over the many-headed monster like a good husbandman, fostering and cultivating the gentle qualities, and preventing the wild ones from growing; he should be making the lion-heart his ally, and in common care of them all should be uniting the several parts with one another and with himself.

Yes, he said, that is quite what the maintainer of justice says.

And so from every point of view, whether of pleasure, honour, or advantage, the approver of justice is right and speaks the truth, and the disapprover is wrong and false and ignorant?

Yes, from every point of view.

Come, now, and let us gently reason with the unjust, who is not intentionally in error. 'Sweet Sir,' we will say to him, 'what think you of things esteemed noble and ignoble? Is not the noble that which subjects the beast to the man, or rather to the god in man; and the ignoble that which subjects the man to the beast?' He can hardly avoid saying Yes – can he now?

Not if he has any regard for my opinion.

But, if he agree so far, we may ask him to answer another question: 'Then how would a man profit if he received gold and silver on the condition that he was to enslave the noblest part of him to the worst? Who can imagine that a man who sold his son or daughter into slavery for money, especially if he sold them into the hands of fierce and evil men, would be the gainer, however large might be the sum which he received? And will anyone say that he is not a miserable caitiff who remorselessly sells his own divine being to that which is most godless and detestable? [590] Eriphyle took the necklace as the price of her husband's life, but he is taking a bribe in order to compass a worse ruin.'

Yes, said Glaucon, far worse – I will answer for him.

Has not the intemperate been censured of old, because in him the huge multiform monster is allowed to be too much at large?

Clearly.

And men are blamed for pride and bad temper when the lion and serpent element in them disproportionately grows and gains strength?

Yes.

And luxury and softness are blamed, because they relax and weaken this same creature, and make a coward of him?

Very true.

And is not a man reproached for flattery and meanness who subordinates the spirited animal to the unruly monster, and, for the sake of money, of which he can never have enough, habituates him in the days of his youth to be trampled in the mire, and from being a lion to become a monkey?

True, he said.

And why are mean employments and manual arts a reproach? Only because they imply a natural weakness of the higher principle; the individual is unable to control the creatures within him, but has to court them, and his great study is how to flatter them.

Such appears to be the reason.

And therefore, being desirous of placing him under a rule like that of the best, we say that he ought to be the servant of the best, in whom the Divine rules; not, as Thrasymachus supposed, to the injury of the servant, but because everyone had better be ruled by divine wisdom dwelling within him; or, if this be impossible, then by an external authority, in order that we may be all, as far as possible, under the same government, friends and equals.

True, he said.

And this is clearly seen to be the intention of the law, which is the ally of the whole city; and is seen also in the authority which we exercise over children, and the refusal to let them be free until we have established in them a principle analogous to the [591] constitution of a state, and by cultivation of this higher element have set up in their hearts a guardian and ruler like our own, and when this is done they may go their ways.

Yes, he said, the purpose of the law is manifest.

From what point of view, then, and on what ground can we say that a man is profited by injustice or intemperance or other baseness, which will make him a worse man, even though he acquire money or power by his wickedness?

From no point of view at all.

What shall he profit, if his injustice be undetected and unpunished? He who is undetected only gets worse, whereas he who is detected and punished has the brutal part of his nature silenced and humanized; the gentler element in him is liberated, and his whole soul is perfected and ennobled by the acquirement of justice and temperance and wisdom, more than the body ever is by receiving gifts of beauty, strength and health, in proportion as the soul is more honourable than the body.

Certainly, he said.

To this nobler purpose the man of understanding will devote the energies of his life. And in the first place, he will honour studies which impress these qualities on his soul and will disregard others?

Clearly, he said.

In the next place, he will regulate his bodily habit and training, and so far will he be from yielding to brutal and irrational pleasures, that he will regard even health as quite a secondary matter; his first object will be not that he may be fair or strong or well, unless he is likely thereby to gain temperance, but he will always desire so to attemper the body as to preserve the harmony of the soul?

Certainly he will, if he has true music in him.

And in the acquisition of wealth there is a principle of order and harmony which he will also observe; he will not allow himself to be dazzled by the foolish applause of the world, and heap up riches to his own infinite harm?

Certainly not, he said.

He will look at the city which is within him, and take heed that no disorder occur in it, such as might arise either from superfluity or from want; and upon this principle he will regulate his property and gain or spend according to his means.

Very true.

[592] And, for the same reason, he will gladly accept and enjoy such honours as he deems likely to make him a better man; but those, whether private or public, which are likely to disorder his life, he will avoid?

Then, if that is his motive, he will not be a statesman.

By the dog, he will! In the city which is his own he certainly will, though in the land of his birth perhaps not, unless he have a divine call.

I understand; you mean that he will be a ruler in the city of which we are the founders, and which exists in idea only; for I do not believe that there is such a one anywhere on earth?

In heaven, I replied, there is laid up a pattern of it, methinks, which he who desires may behold, and beholding, may set his own house in order. But whether such a one exists, or ever will exist in fact, is no matter; for he will live after the manner of that city, having nothing to do with any other.

I think so, he said.

BOOK X

[595a-610e has first reaffirmed the condemnation of mimetic of poetry, resuming the theme from Books II-III, arguing that, like painting, it is at the third remove from the truth, picturing reality by imitation of things that are themselves shadows of the real world. It also encourages the growth of our emotional responses at the expense of reason. This damages that part of us that is immortal, the soul. Its immortality is proved by the principle that each transient thing is destroyed by its own particular evil. But the evil of the soul is vice, and vice itself does not go on to destroy it.]

If the inherent natural vice or evil of the soul is unable to kill or destroy her, hardly will that which is appointed to be the destruction of some other body, destroy a soul or anything else except that of which it was appointed to be the destruction.

Yes, that can hardly be.

But the soul which cannot be destroyed by an evil, whether inherent [611] or external, must exist forever, and if existing forever, must be immortal?

Certainly.

That is the conclusion, I said; and, if a true conclusion, then the souls must always be the same, for if none be destroyed they will not diminish in number. Neither will they increase, for the increase of the immortal natures must come from something mortal, and all things would thus end in immortality.

Very true.

But this we cannot believe – reason will not allow us – any more than we can believe the soul, in her truest nature, to be full of variety and difference and dissimilarity.

What do you mean? he said.

The soul, I said, being, as is now proven, immortal, must be the fairest of compositions and cannot be compounded of many elements?

Certainly not.

Her immortality is demonstrated by the previous argument, and there are many other proofs; but to see her as she really is, not as we now behold her, marred by communion with the body and other miseries, you must contemplate her with the eye of reason, in her original purity; and then her beauty will be revealed, and justice and injustice and all the things which we have described will be manifested more clearly. Thus far, we have spoken the truth concerning her as she appears at present, but we must remember also that we have seen her only in a condition which may be compared to that of the sea-god

Glaucus, whose original image can hardly be discerned because his natural members are broken off and crushed and damaged by the waves in all sorts of ways, and incrustations have grown over them of seaweed and shells and stones, so that he is more like some monster than he is to his own natural form. And the soul which we behold is in a similar condition, disfigured by ten thousand ills. But not there, Glaucon, not there must we look.

Where then?

At her love of wisdom. Let us see whom she affects, and what society and converse she seeks in virtue of her near kindred with the immortal and eternal and divine; also how different she would become if wholly following this superior principle, and borne by a divine impulse out of the ocean in which she now is, and disengaged from the stones and shells and things of earth and rock which in wild variety spring up [612] around her because she feeds upon earth, and is overgrown by the good things of this life as they are termed: then you would see her as she is, and know whether she have one shape only or many, or what her nature is. Of her affections and of the forms which she takes in this present life I think that we have now said enough.

True, he replied.

And thus, I said, we have fulfilled the conditions of the argument; we have not introduced the rewards and glories of justice, which, as you were saying, are to be found in Homer and Hesiod; but justice in her own nature has been shown to be best for the soul in her own nature. Let a man do what is just, whether he have the ring of Gyges or not, and even if in addition to the ring of Gyges he put on the helmet of Hades.

Very true.

And now, Glaucon, there will be no harm in further enumerating how many and how great are the rewards which justice and the other virtues procure to the soul from gods and men, both in life and after death.

Certainly not, he said.

Will you repay me, then, what you borrowed in the argument?

What did I borrow?

The assumption that the just man should appear unjust and the unjust just: for you were of opinion that even if the true state of the case could not possibly escape the eyes of gods and men, still this admission ought to be made for the sake of the argument, in order that pure justice might be weighed against pure injustice. Do you remember?

I should be much to blame if I had forgotten.

Then, as the cause is decided, I demand on behalf of justice that the estimation in which she is held by gods and men and which we acknowledge to be her due should now be restored to her by us; since she has been shown to confer reality, and not to deceive those who truly possess her, let what has been taken from her be given back, that so she may win that palm of appearance which is hers also, and which she gives to her own.

The demand, he said, is just.

In the first place, I said – and this is the first thing which you will have to give back – the nature both of the just and unjust is truly known to the gods.

Granted.

And if they are both known to them, one must be the friend and the other the enemy of the gods, as we admitted from the beginning?

True.

And the friend of the gods may be supposed to receive from them all things at their best, [613] excepting only such evil as is the necessary consequence of former sins?

Certainly.

Then this must be our notion of the just man, that even when he is in poverty or sickness, or any other seeming misfortune, all things will in the end work together for good to him in life and death: for the gods have a care of anyone whose desire is to become just and to be like God, as far as man can attain the divine likeness, by the pursuit of virtue?

Yes, he said; if he is like God he will surely not be neglected by him.

And of the unjust may not the opposite be supposed?

Certainly.

Such, then, are the palms of victory which the gods give the just?

That is my conviction.

And what do they receive of men? Look at things as they really are, and you will see that the clever unjust are in the case of runners, who run well from the starting-place to the goal but not back again from the goal: they go off at a great pace, but in the end only look foolish, slinking away with their ears dragging on their shoulders, and without a crown; but the true runner comes to the finish and receives the prize and is crowned. And this is the way with the just; he who endures to the end of every action and occasion of his entire life has a good report and carries off the prize which men have to bestow.

True.

And now you must allow me to repeat of the just the blessings which you were attributing to the fortunate unjust. I shall say of them, what you were saying of the others, that as they grow older, they become rulers in their own city if they care to be; they marry whom they like and give in marriage to whom they will; all that you said of the others I now say of these. And, on the other hand, of the unjust I say that the greater number, even though they escape in their youth, are found out at last and look foolish at the end of their course, and when they come to be old and miserable are flouted alike by stranger and citizen; they are beaten and then come those things unfit for ears polite, as you truly term them; they will be racked and have their eyes burned out, as you were saying. And you may suppose that I have repeated the remainder of your tale of horrors. But will you let me assume, without reciting them, that these things are true?

Certainly, he said, what you say is true.

These, then, are the prizes and rewards and gifts which are bestowed upon the just by gods and men in this present life, [614] in addition to the other good things which justice of herself provides.

Yes, he said; and they are fair and lasting.

And yet, I said, all these are as nothing either in number or greatness in comparison with those other recompenses which await both just and unjust after death. And you ought to hear them, and then both just and unjust will have received from us a full payment of the debt which the argument owes to them.

Speak, he said; there are few things which I would more gladly hear.

Well, I said, I will tell you a tale; not one of the tales which Odysseus tells to the hero Alcinous, yet this too is a tale of a hero, Er the son of Armenius, a Pamphylian by birth. He was slain in battle, and ten days afterwards, when the bodies of the dead were taken up already in a state of corruption, his body was found unaffected by decay, and carried away home to be buried. And on the twelfth day, as he was lying on the funeral pile, he returned to life and told them what he had seen in the other world. He said that when his soul left the body he went on a journey with a great company, and that they came to a mysterious place at which there were two openings in the earth; they were near together, and over against them were two other openings in the heaven above. In the intermediate space there were judges seated, who commanded the just, after they had given judgment on them and had bound their sentences in front of them, to ascend by the heavenly way on the right hand; and in like manner the unjust were bidden by them to descend by the lower way on the left hand; these also bore the symbols of their deeds, but fastened on their backs. He drew near, and they told him that he was to be the messenger who would carry the report of the other world to men, and they bade him hear and see all that was to be heard and seen in that place. Then he beheld and saw on one side the souls departing at either opening of heaven and earth when sentence had been given on them; and at the two other openings other souls, some ascending out of the earth dusty and worn with travel, some descending out of heaven clean and bright. And arriving ever and anon they seemed to have come from a long journey, and they went forth with gladness into the meadow, where they encamped as at a festival; and those who knew one another embraced and conversed, the souls which came from earth curiously enquiring about the things above, and the souls which came from heaven about the things beneath. And they told one another of what had happened by the way, those from below weeping and [615] sorrowing at the remembrance of the things which they had endured and seen in their journey beneath the earth (now the journey lasted a thousand years), while those from above were describing heavenly delights and visions of inconceivable beauty. The story, Glaucon, would take too long to tell; but the sum was this: – He said that for every wrong which they had done to anyone they suffered tenfold; or once in a hundred years – such being reckoned to be the length of man's life, and the penalty being thus paid ten times in a thousand years. If, for example, there were any who had been the cause of many deaths, or had betrayed or enslaved cities or armies, or been guilty of any other evil behaviour, for each and all of their offences they received punishment ten times over, and the rewards of beneficence and justice and holiness were in the same proportion. I need hardly repeat what he said concerning young children dying almost as soon as they were born. Of piety and impiety to gods and parents, and of murderers, there were retributions other and greater far which he described. He mentioned that he was present when one of the spirits asked another, 'Where is Ardiaeus the Great?' (Now this Ardiaeus lived a thousand years before the time of Er: he had been the tyrant of some city of Pamphylia, and had murdered his aged father and his elder brother, and was said to have committed many other abominable crimes.) The answer of the other spirit was: 'He comes not hither and will never come. And this,' said he, 'was one of the dreadful sights which we ourselves witnessed. We were at the mouth of the cavern, and, having completed all our experiences, were about to reascend, when of a sudden Ardiaeus appeared and several others, most of whom were tyrants; and there were also besides the tyrants private individuals who had been great criminals: they were just, as they fancied, about to return into the upper world, but the mouth, instead of admitting them, gave a roar, whenever any of these incurable sinners or someone who had not been sufficiently punished tried to ascend; and then wild men of fiery aspect, who were standing by and heard the sound, seized and carried them off; and Ardiaeus and others they bound [616] head and foot and hand, and threw them down and flayed them with scourges, and dragged them along the road at the side, carding them on thorns like wool, and declaring to the passers-by what were their crimes, and that they were being taken away to be cast into hell.' And of all the many terrors which they had endured, he said that there was none like the terror which each of them felt at that moment, lest they should hear the voice; and when there was silence, one by one they ascended with exceeding joy. These, said Er, were the penalties and retributions, and there were blessings as great.

Now when the spirits which were in the meadow had tarried seven days, on the eighth they were obliged to proceed on their journey, and, on the fourth day after, he said that they came to a place where they could see from above a line of light, straight as a column, extending right through the whole heaven and through the earth, in colour resembling the rainbow, only brighter and purer; another day's journey brought them to the place, and there, in the midst of the light, they saw the ends of the chains of heaven let down from above: for this light is the belt of heaven, and holds together the circle of the universe, like the under-girders of a trireme. From these ends is extended the spindle of Necessity, on which all the revolutions turn. The shaft and hook of this spindle are made of steel, and the whorl is made partly of steel and also partly of other materials. Now the whorl is in form like the whorl used on earth; and the description of it implied that there is one large hollow whorl which is quite scooped out, and into this is fitted another lesser one, and another, and another, and four others, making eight in all, like vessels which fit into one another; the whorls show their edges on the upper side, and on their lower side all together form one continuous whorl. This is pierced by the spindle, which is driven home through the centre of the eighth. The first and outermost whorl has the rim broadest, and the seven inner whorls are narrower, in the following proportions – the sixth is next to the first in size, the fourth next to the sixth; then comes the eighth; the seventh is fifth, the fifth is sixth, the third is seventh, last and eighth comes the second. The largest (or fixed stars) is spangled, and the seventh (or sun) is brightest; the eighth (or moon) coloured by the reflected light of the seventh; [617] the second and fifth (Saturn and Mercury) are in colour like one another, and yellower than the preceding; the third (Venus) has the whitest light; the fourth (Mars) is reddish; the sixth (Jupiter) is in whiteness second. Now the whole spindle has the same motion; but, as the whole revolves in one direction, the seven inner circles move slowly in the other, and of these the swiftest is the eighth; next in swiftness are the seventh, sixth, and fifth, which move together; third in swiftness appeared to move according to the law of this reversed motion the fourth; the third appeared fourth and the second fifth. The spindle turns on the knees of Necessity; and on the upper surface of each circle is a siren, who goes round with them, hymning a single tone or note. The eight together form one harmony; and round about, at equal intervals, there is another band, three in number, each sitting upon her throne: these are the Fates, daughters of Necessity, who are clothed in white robes and have chaplets upon their heads, Lachesis and Clotho and Atropos, who accompany with their voices the harmony of the sirens – Lachesis singing of the past, Clotho of the present, Atropos of the future; Clotho from time to time assisting with a touch of her right hand the revolution of the outer circle of the whorl or spindle, and Atropos with her left hand touching and guiding the inner ones, and Lachesis laying hold of either in turn, first with one hand and then with the other.

When Er and the spirits arrived, their duty was to go at once to Lachesis; but first of all there came a prophet who arranged them in order; then he took from the knees of Lachesis lots and samples of lives, and having mounted a high pulpit, spoke as follows: 'Hear the word of Lachesis, the daughter of Necessity. Mortal souls, behold a new cycle of life and mortality. Your genius will not be allotted to you, but you will choose your genius; and let him who draws the first lot have the first choice, and the life which he chooses shall be his destiny. Virtue is free, and as a man honours or dishonours her he will have more or less of her; the responsibility is with the chooser – God is justified.' When the Interpreter had thus spoken he scattered lots indifferently among them all, and each of them took up the lot which fell near him, all but Er himself (he was not allowed), [618] and each as he took his lot perceived the number which he had obtained. Then the Interpreter placed on the ground before them the samples of lives; and there were many more lives than the souls present, and they were of all sorts. There were lives of every animal and of man in every condition. And there were tyrannies among them, some lasting out the tyrant's life, others which broke off in the middle and came to an end in poverty and exile and beggary; and there were lives of famous men, some who were famous for their form and beauty as well as for their strength and success in games, or, again, for their birth and the qualities of their ancestors; and some who were the reverse of famous for the opposite qualities. And of women likewise; there was not, however, any definite character in them, because the soul, when choosing a new life, must of necessity become different. But there was every other quality, and they all mingled with one another, and also with elements of wealth and poverty, and disease and health; and there were mean states also. And here, my dear Glaucon, is the supreme peril of our human state; and therefore the utmost care should be taken. Let each one of us leave every other kind of knowledge and seek and follow one thing only, if peradventure he may be able to learn and may find someone who will make him able to learn and discern between good and evil, and so to choose always and everywhere the better life as he has opportunity. He should consider the bearing of all these things which have been mentioned severally and collectively upon virtue; he should know what the effect of beauty is when combined with poverty or wealth in a particular soul, and what are the good and evil consequences of noble and humble birth, of private and public station, of strength and weakness, of cleverness and dullness, and of all the natural and acquired gifts of the soul, and the operation of them when conjoined; he will then look at the nature of the soul, and from the consideration of all these qualities he will be able to determine which is the better and which is the worse; and so he will choose, giving the name of evil to the life which will make his soul more unjust, and good to the life which will make his soul more just; all else he will disregard. For we have seen and know that this is the best choice both in life and after death. [619] A man must take with him into the world below an adamantine faith in truth and right, that there too he may be undazzled by the desire of wealth or the other allurements of evil, lest, coming upon tyrannies and similar villainies, he do irremediable wrongs to others and suffer yet worse himself; but let him know how to choose the mean and avoid the extremes on either side, as far as possible, not only in this life but in all that which is to come. For this is the way of happiness.

And according to the report of the messenger from the other world this was what the prophet said at the time: 'Even for the last comer, if he chooses wisely and will live diligently, there is appointed a happy and not undesirable existence. Let not him who chooses first be careless, and let not the last despair.' And when he had spoken, he who had the first choice came forward and in a moment chose the greatest tyranny; his mind having been darkened by folly and sensuality, he had not thought out the whole matter before he chose, and did not at first sight perceive that he was fated, among other evils, to devour his own children. But when he had time to reflect, and saw what was in the lot, he began to beat his breast and lament over his choice, forgetting the proclamation of the prophet; for, instead of throwing the blame of his misfortune on himself, he accused chance and the gods, and everything rather than himself. Now he was one of those who came from heaven, and in a former life had dwelt in a well-ordered State, but his virtue was a matter of habit only, and he had no philosophy. And it was true of others who were similarly overtaken, that the greater number of them came from heaven and therefore they had never been schooled by trial, whereas the pilgrims who came from earth having themselves suffered and seen others suffer, were not in a hurry to choose. And owing to this inexperience of theirs, and also because the lot was a chance, many of the souls exchanged a good destiny for an evil or an evil for a good. For if a man had always on his arrival in this world dedicated himself from the first to sound philosophy, and had been moderately fortunate in the number of the lot, he might, as the messenger reported, be happy here, and also his journey to another life and return to this, instead of being rough and underground, would be smooth and heavenly. Most curious, he said, was the spectacle – [620] sad and laughable and strange; for the choice of the souls was in most cases based on their experience of a previous life. There he saw the soul which had once been Orpheus choosing the life of a swan out of enmity to the race of women, hating to be born of a woman because they had been his murderers; he beheld also the soul of Thamyras choosing the life of a nightingale; birds, on the other hand, like the swan and other musicians, wanting to be men. The soul which obtained the twentieth lot chose the life of a lion, and this was the soul of Ajax the son of Telamon, who would not be a man, remembering the injustice which was done him in the judgment about the arms. The next was Agamemnon, who took the life of an eagle, because, like Ajax, he hated human nature by reason of his sufferings. About the middle came the lot of Atalanta; she, seeing the great fame of an athlete, was unable to resist the temptation: and after her there followed the soul of Epeus the son of Panopeus passing into the nature of a woman cunning in the arts; and far away among the last who chose, the soul of the jester Thersites was putting on the form of a monkey. There came also the soul of Odysseus having yet to make a choice, and his lot happened to be the last of them all. Now the recollection of former toils had disenchanted him of ambition, and he went about for a considerable time in search of the life of a private man who had no cares; he had some difficulty in finding this, which was lying about and had been neglected by everybody else; and when he saw it, he said that he would have done the same had his lot been first instead of last, and that he was delighted to have it. And not only did men pass into animals, but I must also mention that there were animals tame and wild who changed into one another and into corresponding human natures – the good into the gentle and the evil into the savage, in all sorts of combinations.

All the souls had now chosen their lives, and they went in the order of their choice to Lachesis, who sent with them the genius whom they had severally chosen, to be the guardian of their lives and the fulfiller of the choice: this genius led the souls first to Clotho, and drew them within the revolution of the spindle impelled by her hand, thus ratifying the destiny of each; and then, when they were fastened to this, carried them to Atropos, who spun the threads and made them irreversible, whence without turning round they passed beneath [621] the throne of Necessity; and when they had all passed, they marched on in a scorching heat to the plain of Forgetfulness, which was a barren waste destitute of trees and verdure; and then towards evening they encamped by the river of Unmindfulness, whose water no vessel can hold; of this they were all obliged to drink a certain quantity, and those who were not saved by wisdom drank more than was necessary; and each one as he drank forgot all things. Now after they had gone to rest, about the middle of the night there was a thunderstorm and earthquake, and then in an instant they were driven upwards in all manner of ways to their birth, like stars shooting. He himself was hindered from drinking the water. But in what manner or by what means he returned to the body he could not say; only, in the morning, awaking suddenly, he found himself lying on the pyre.

And thus, Glaucon, the tale has been saved and has not perished, and will save us if we are obedient to the word spoken; and we shall pass safely over the river of Forgetfulness and our soul will not be defiled. Wherefore my counsel is, that we hold fast ever to the heavenly way and follow after justice and virtue always, considering that the soul is immortal and able to endure every sort of good and every sort of evil. Thus shall we live dear to one another and to the gods, both while remaining here and when, like conquerors in the games who go round to gather gifts, we receive our reward. And it shall be well with us both in this life and in the pilgrimage of a thousand years which we have been describing.

Biographies

Plato

Republic

Among the founders of western thought, Plato (*c.* 428–348 BCE) came from Athens, where he established the original Academy. His most celebrated student was Aristotle (348–22 BCE). So central were intellectual exchange and questioning for Plato that he wrote his works as dialogues, one voice being that of his own true-life tutor Socrates (c. 470–399 BCE).

Benjamin Jowett

Translator: The Republic of Plato

Benjamin Jowett (1817–93) was a translator, classical scholar, English tutor and Anglican clergyman. Born in London, Jowett received a scholarship at age 18 to attend Balliol College, Oxford, where his interest in classical studies led to his recognition as a Fellow in 1838 while still an undergraduate. For many years afterward he served as Master of the college. He is best known for his translations of works by Plato and the Athenian historian and general Thucydides (*c.* 460–400 BCE).

Professor Harold Tarrant

New Introduction & Abridgement

Harold Tarrant studied at Cambridge and Durham Universities, and worked at the University of Sydney from 1973 to 1993, when he took up the Chair of Classics at the University of Newcastle, Australia. Though now retired, he retains his Newcastle affiliation as Emeritus Professor. His principle focus is ancient Platonism, from Plato to Olympiodorus. He has authored or co-authored ten books in that field, and edited or co-edited others.

Professor Marika Taylor

Series Foreword

Professor Marika Taylor is a Professor of Theoretical Physics and Head of School within Mathematical Sciences at the University of Southampton. Her research interests include all aspects of string theory, gravitational physics and quantum field theory. In recent years much of her work has been focused on holographic dualities and their implications. Marika's research has featured in such publications as *Physical Review*, *Journal of High Energy Physics* and *General Relativity and Gravitation* among others.